CRITICAL THINKING, SCIENCE, AND PSEUDOSCIENCE

Caleb W. Lack, PhD, is an associate professor of psychology at the University of Central Oklahoma. He is the author or editor of five prior books: *Tornadoes, Children, and Posttraumatic Stress; Anxiety Disorders: An Introduction; Mood Disorders: An Introduction; Psychology Gone Astray*; and *Obsessive-Compulsive Disorder: Etiology, Phenomenology, & Treatment*. In addition, he has authored over 40 peer-reviewed journal articles and book chapters. He received his doctorate in clinical psychology from Oklahoma State University and has won numerous awards for his innovative teaching and research. Outside of the realm of clinical psychology, he also teaches undergraduate and graduate courses on critical thinking, science, and pseudoscience. He writes the "Great Plains Skeptic" column on the Skeptic Ink Network and presents frequently on how to think critically about paranormal and supernatural claims. You can learn more by visiting his website at www.caleblack.com.

Jacques A. Rousseau, MA, is currently a lecturer in the School of Management Studies at the University of Cape Town (UCT). He studied philosophy and English at UCT between 1992 and 1996, before beginning an academic career that has involved teaching multiple courses in critical thinking, business ethics, moral philosophy, and sociology. His primary research areas involve epistemic standards in journalism and public media, addiction, critical thinking, and religious conflict. He sits on various governance bodies at UCT, including its Council, and is the chair of the Academic Freedom Committee. In 2009, he founded (and currently chairs) the Free Society Institute, a nonprofit organization dedicated to promoting free thought, free speech, secular humanism, and scientific reasoning. You can learn more by visiting his website at www.jacquesrousseau.com.

CRITICAL THINKING, SCIENCE, AND PSEUDOSCIENCE

Why We Can't Trust Our Brains

Caleb W. Lack, PhD
Jacques Rousseau, MA

SPRINGER PUBLISHING COMPANY
NEW YORK

Springer Publishing Company, LLC
11 West 42nd Street
New York, NY 10036
www.springerpub.com

Acquisitions Editor: Nancy Hale
Composition: diacriTech

ISBN: 978-0-8261-9419-0
e-book ISBN: 978-0-8261-9426-8

Instructors' Materials: Qualified instructors may request supplements by e-mailing textbook@springerpub.com:
Instructor's Manual ISBN: 978-0-8261-9432-9
PowerPoint ISBN: 978-0-8261-9438-9

16 17 18 19 20 / 5 4 3 2 1

Library of Congress Cataloging-in-Publication Data
Names: Lack, Caleb W., 1978- , author. | Rousseau, Jacques (Jacques André), 1971- , author.
Title: Critical thinking, science, and pseudoscience : why we can't trust our brains / Caleb W. Lack, Jacques Rousseau.
Description: New York, NY : Springer Publishing Company, LLC, [2016] | Includes bibliographical references and index.
Identifiers: LCCN 2015040040| ISBN 9780826194190 | ISBN 9780826194268 (e-book)
Subjects: | MESH: Thinking. | Health Knowledge, Attitudes, Practice. | Religion and Science.
Classification: LCC BF441 | NLM BF 441 | DDC 153.4/2--dc23 LC record available at http://lccn.loc.gov/2015040040

Printed in the United States of America by Bradford and Bigelow.

For my son, Lucian. I hope that the world you grow up in contains at least a few more critical thinkers in it because of our work.

—CWL

To my father, Jacques Rossouw—a model of scientific integrity and a lifelong inspiration for careful and critical thinking. Thank you.

—JR

CONTENTS

FOREWORD: BRAINS, HEARTS, GUTS, AND GENITALS

Ask an actor, "How do you connect with an audience?" and you're likely to hear that he or she will aim at an audience on four levels. An actor can appeal to the intellect, to emotions, to instinct, or, shall we say genteelly, to animal urges. In other words, brains, hearts, guts, and genitals.

Guess which will be the first choice?

Right. The farther down the body you go, the easier, quicker, and more effective is the connection to an audience. You can see this in magazine advertising or on Internet social media. "Click-bait" focuses heavily on guts and genitals—and all those cute cat and puppy videos are an appeal to the heart. You don't see a lot of click bait appealing to brains. Twitter and YouTube are not the first places you turn to for logical, intellectual discourse.

So why did the authors of this book focus at the top of the list, rather than somewhere lower down?

Because even though the downward end of the scale of brains, hearts, guts, and genitals works in acting, YouTube posts, advertising, and other forms of communication, it's not necessarily how you want to live your life. As the starring actor in your own life's drama, you are going to be much better off if you focus on the top of that scale for most of the decisions in your life.

It's not going out on a limb to say that you *really* should start at the top when it's a big decision. If you know how to use your brain to think critically about the world around you, most of the time you're going to come out ahead. Arguably, relying on the topmost anatomical features (head and heart) will get you farther than guts and genitals.

I'm not proposing that you give up heart, guts, and genitals; you're not likely to comply, and really, it's not necessary anyhow. If you're walking to class or work, say, and you put in your earbuds to listen to some music, you're not likely to need to employ a lot of the critical thinking skills so usefully outlined in this book: you're going to choose your selections depending on how you feel. Are you feeling really mellow? Are you feeling *too* mellow, and need to get your brain firing before you get there? Okay, crank up something with a great beat, or whatever works for you to get your heart rate up. We make decisions all the time that don't involve a lot of logic or empirical evidence.

But let's say you get a phone call from a lawyer in a far-off foreign country who has really bad news about a friend or relative of yours who has just been arrested on a trumped-up drug charge and needs $100 to bribe his way out of jail. You need to buy some prepaid debit cards, and read the serial numbers to the lawyer so he can get the money to rescue your friend. Whatever you do, the lawyer warns, don't tell anyone, even other friends, because the situation is so perilous, and the bribe could fall through if the authorities get wind of it. And do it right away because even a few hours in the jails of this country are a nightmare for your poor friend.

Here's where you need a little critical thinking. Of course your first reaction will be to want to help a loved one. Hearts are good things, but don't rely on them for analysis. For circumstances like this, you need to use your brain, not your heart. Every year, scammers obtain hundreds of thousands of (untraceable) dollars from well-meaning people who believe they are doing the right thing for a loved one. It's important to have the tools to think critically about such circumstances and to evaluate such claims.

Going with your gut also has its shortcomings. When you "just know" something, you might need to stop and think whether your belief comes from a reliable source or is the product of stereotypes, inadequate or mistaken information, or even superstition. It's especially worrisome when people in decision-making positions cede thinking to their instincts. President George W. Bush famously "looked [Russian President Vladimir Putin] in the eye" and "found him trustworthy"; we would hope that successful international diplomacy would rely on a more systematic procedure for assessing the trustworthiness of decision makers.

Heads, hearts, guts … we've now made our way to the last rung of the ladder: genitals.

Is it pretty obvious why you shouldn't make major decisions based on your sexual impulses? I thought so.

But lest you think that I'm advocating a really boring life, where every decision is the product of critical thinking and rationality, à la Mr. Spock of *Star Trek*, let me reassure you. I'm not. Life is a balance, and life would be pretty dull if we didn't have love, and wonder, and excitement, and awe, and delight, and lots of other things that don't have to be the product of logic and reason.

There's a famous etching by Francisco Goya portraying the artist slumped at his desk in exhausted sleep, while above him fly scary-looking bats and owls—symbols of evil in Goya's eighteenth-century Spain. The title is "The Sleep of Reason Produces Monsters," and it has become something of a slogan for scientific skeptics and other rationalists who worry about what happens when we don't use our brains.

Yet it's useful to look at the full epigraph to the etching, which reads, in translation, "Imagination abandoned by reason produces impossible monsters: united with her, she is the mother of the arts and the source of their wonders."

To be "wonderful"—full of wonder—the arts require not only imagination, but also reason, says Goya. I'd like to think of this as a metaphor. Hearts, guts, and genitals need to be guided by reason, lest they result in bad decisions—Goya's "monsters." We may fall prey to scammers who cost us—whether money, time, emotions, or something else we hold dear. Even sincere people may mislead us into believing things that are not true, or not believing things that are true, and bad decisions and consequent suffering can take place. Without critical thinking, we may be led to accept false ideas about health, finances, personal relationships, or other issues that can injure us or our loved ones.

No one can, or should, abandon brains, or heart, or guts, or genitals. We need all of these to be complete human beings. But Goya's point is to not let your reason sleep! Only *with* reason—engaged, not relegated to the sidelines, not sleeping—can we ensure that our other human qualities won't lead us astray.

Enjoy all the parts of your body! And read this book to learn how to be sure that reason isn't sleeping; brains are at the top, after all!

Eugenie C. Scott, PhD
Founding Executive Director
National Center for Science Education
Oakland, California

PREFACE

Critical Thinking, Science, and Pseudoscience: Why We Can't Trust Our Brains offers an introduction to various issues that relate to our ability to navigate the world of information effectively and efficiently. The first obvious issue that arises when trying to make sense out of all the noise we're subjected to is how arguments and evidence work—summarized here as the skill of critical thinking. But it's not enough to know how to think well in theory, because in practice—as you no doubt know all too well—we often don't have the time or ability to be as careful with our reasoning as we might like to be. We're bombarded with conflicting information, and in an effort to save time and make decisions easier, we've all developed various strategies for coping with information overload. Yet these strategies can sometimes lead us astray, so it's important to talk about how humans actually think—not just how we should think—and to also explore the various ways in which we might be led to confusion not only by our own brains, but also by the misleading or inaccurate information we're sometimes exposed to.

This book is unique in two important ways: The first is the broad cross-disciplinary approach that it brings to the topic of critical thinking, and the second is that we take care, even when discussing ideas that we regard as deeply flawed, to not belittle or ridicule those who hold such views. Being a consistently critical thinker is not easy for anyone, and is not aided by mockery or misunderstanding. But more to the point, the authors recognize that we are all prone to errors of similar sorts, and think that dialogue and understanding, rather than caricature and hectoring, are more productive tools for minimizing those errors.

On the first aspect, the cross-disciplinary approach: Although the book takes a primarily psychological and philosophical look at a diverse range of topics, we also incorporate the perspectives of biology, physics, medicine, history, and so on. As an example, when discussing alien abductions, we discuss not only the historical perspective on why we have the image of aliens being thin, gray-skinned, big-eyed humanoids, but also the problems with eyewitness testimony and memory. When discussing cryptids ("hidden" creatures such as the Loch Ness Monster or Bigfoot), we discuss not only the problems in terms of plausibility and evidence, but also the problems with the existence of such creatures from a biological perspective, such as food supplies and lack of physical evidence. This emphasis on examining phenomena and claims from multiple perspectives is intended to help the reader consider different levels of (rational) explanation that lead to the same conclusion—in this case, that pseudoscience is bunk—while also offering intellectual resources that can be generalized for reaching that same conclusion for topic areas the book does not address.

In terms of organization and approach, the second unique aspect of this book is expressed in a Michael Shermer quote inspired by Baruch Spinoza: "I have made a ceaseless effort not to ridicule, not to bewail, not to scorn human actions, but to understand them" (Shermer, 2002, p. 41). As such, the book will not just focus on "debunking" various pseudoscientific topics, but also will help the reader understand why people believe in such things. We begin by discussing the need for critical thinking (Chapter 1), centering around the question of "What's the harm?" and showing exactly how not thinking critically can cause minor and major damage to both individuals and society. We then operationally define *science* (Chapter 2), and show how it is similar to and different from other types of knowledge and understanding. Chapter 3 describes pseudoscience, illustrates some of the common mistakes made in pseudoscientific thinking, and lays some of the groundwork for discussions of particular examples of pseudoscience in later chapters. We then move to a discussion of the key principles of critical thinking in Chapter 4, before spending the next two chapters examining why it is so difficult to be a critical thinker. These discussions are framed in terms of the questions "Why can't we trust our brains?" (Chapter 5) and "Why can't we trust our world?" (Chapter 6). In Chapter 5, we focus on how the human brain does *not* process information in a rational, logical fashion and instead is rife with natural biases. In Chapter 6, we expose many of the social factors that come into play that prevent one from gaining an unbiased, critical perspective on information.

After laying the foundations of how we would ideally think about things, but then showing how our brains and world complicate that ideal, we introduce specific pseudoscience topics as illustrative examples of this tension, applying the principles of critical thinking to examine the veracity of various claims. Chapter 7 discusses alien visitation and abduction. Chapter 8 tackles the subject of psychic powers and those who claim to have them. Chapter 9 deals with some of the strange creatures some people claim to see, or believe that they might have seen—Bigfoot, the Loch Ness monster, and so on.

Chapters 10 through 12 combine as an extensive look at alternative medicine. Chapter 10 lays the groundwork in discussing what alternative medicine is and when medical treatments should or should not be considered trustworthy. The next two chapters examine two areas where alternative medicine is prevalent—physical health (chiropractic treatment, acupuncture, etc.) and mental health (e.g., facilitated communication for autism and sensory integration therapy). All of this is done through the lens of critical thinking, and includes discussion about *why* people believe in alternative treatments, using the principles described in Chapters 5 and 6.

Chapter 13 takes a look at some of the more popular conspiracy theories, including the Illuminati and the New World Order, "false flag" operations, suppression of technology, and apocalyptic predictions. Our penultimate chapter (Chapter 14) then takes a look at how religion and culture impact science and vice versa, using the narrative of the "culture wars" surrounding topics such as heliocentrism, the theory of evolution via natural selection, climate change, and so on. We then conclude, in Chapter 15, with a restatement of the importance of critical thinking in one's day-to-day life, and provide resources for encouraging critical thinking in our readers' communities.

Although this book will be of great interest to those specifically studying critical thinking—for example, those taking a course on it or studying for a qualification with a primary or secondary focus on critical thinking—the authors are of the firm view that everyone can benefit from becoming a better critical thinker, whether you have to take a course in it or not! So, if you're the sort of person who reads and enjoys popular science books and articles, such as those by the likes of Ben Goldacre, Richard Dawkins, Michael Shermer, Carl Sagan, and Neil DeGrasse Tyson, we think there is a good chance you will enjoy this book also.

While we have referenced key claims, the book is not intended as an academic monograph, and we wouldn't want it to read as such. We hope it will introduce an interested member of the public or an undergraduate student to key issues in how to reason critically and scientifically, and to help this individual to communicate those ideas to others. We hope you enjoy reading it as much as we've enjoyed writing it for you.

Qualified instructors can e-mail textbook@springerpub.com to receive an Instructor's Manual and a chapter-based PowerPoint deck that will aid in engaging and inspiring your students to become better critical thinkers.

REFERENCE

Shermer, M. (2002). *Why people believe weird things: Pseudoscience, superstition, and other confusions of our time*. New York, NY: Holt.

ACKNOWLEDGMENTS

As with any work, this book was not created in a vacuum and was instead shepherded into being via the help of numerous individuals. We would like to first and foremost thank our lovely, long-suffering wives, Alison and Signe, who put up with us writing all hours of the day and night (not to mention just being married to us in general). We would also like to thank our editor at Springer, Nancy S. Hale, for her willingness to take a gamble on a new book and for her guidance through the process. Her assistant editor, Amanda Carswell, also kept us on track and straightened out our various mistakes. Any that remain are solely our fault!

We greatly appreciated all the comments and insights from people who read and commented on the chapters as we were working on them, foremost among them the wonderful Sharon and Jeff Madison. Their careful reading and suggestions certainly improved our book and helped us to tell a solid story about what it means to become a scientific skeptic.

The foreword and afterword were generously written by Dr. Eugenie Scott and Dr. Scott Lilienfeld. Their insight and outstanding work in the field of critical thinking (in biology and psychology, respectively) have been and will continue to be inspirations not only for us but for generations to come.

Thank you to all our students over the years, whose questions and enthusiasm for the subject matter in this book has served as motivation to be better teachers, and also as a useful safeguard against laziness and dogmatic slumber. They also helped us hone our keen senses of humor. If you think our jokes are bad in this book, you should have heard them 10 years ago.

Finally, thanks are due to all our friends inside and outside the skeptical community who have supported us and tolerated our fondness for argument, both online and in real life. This includes, but is not limited to, Ed Clint (who cofounded the Skeptic Ink Network and brought the two of us together), Jean-Pierre Rossouw, Tom Moultrie, Kathryn Stinson, Bryan Gruneberg, Kelly Quantrill, Kent Solomon, James Lennox, David Benatar, Jonathan Pearce, Arpana Dattilo, Matt Korstjens, Damion Reinhardt, Chas Stewart, and Austin Moore.

CHAPTER 1

WHY DO WE NEED CRITICAL THINKING?

Being able to change our minds about something—to realize that we hold a false belief, and to then correct it—is an invaluable human capability. False beliefs can confuse debates, lead us to poorer choices than we might otherwise make, and, in general, limit our opportunities to make the most of our circumstances. Yet, it's a capability that we can sometimes take for granted, so much so that we forget that it can be exercised and strengthened—or that we can become worse at it over time, if we allow faulty assumptions or poor reasoning to become entrenched in our minds. Even if we can recall recent cases in which we've changed our minds about something—maybe the extent to which you favor one political candidate over another, the performance of your sports team, or where you should invest your money—those easy-to-remember cases might obscure an underlying, and perhaps pervasive, tendency not to interrogate our beliefs as often as we should.

Later in this book, you will read about *confirmation bias* (in short, our habit of noticing evidence that confirms something we already believe, and not noticing evidence that complicates the picture or disproves our belief), and this very common bias affects our ability to recognize when we're not *thinking* as well as we could or should—irrespective of *what* we are thinking about (Trout, 2002). So it's not only the case that we might be unduly complacent about particular beliefs, but also (more worrisomely) that we might become unduly complacent about our thinking patterns in general, or about the ways in which information in the world and its presentation could contribute to making that task difficult for us.

The capacity to change our minds, or to approach issues in a way that gives us the best chance of reaching the most justified conclusion, is one that needs nurturing and exercising. If we neglect it, we become worse at doing it—and unfortunately, we're very good at hiding our failures to ourselves. This, in short, is why we need both critical thinking and to exercise our critical-thinking skills: we can benefit from learning how to think things through, or reminding ourselves of the ways in which we let ourselves down in our sincere efforts to do so. It's true that the Internet, and social media in particular, can overwhelm (or at least complicate) our best intentions and efforts. And, contrary to the promise that the Internet holds for robust and informative debate, communication studies indicate that social media, at least, could contribute to a "spiral of silence," in which we mostly end up sharing views only when we think that our audience is likely to agree with us already (Hampton et al., 2014). Later chapters explore the role of social media and the Internet in complicating our thinking and our conversations, but for now, let's focus on a fuller explanation of why we should continue to strive to be better thinkers.

KNOWLEDGE AND EMPOWERMENT

Consider these two statements:

- It's relatively easy to fool someone who doesn't understand the topic under discussion (assuming that you do understand it, or can plausibly pretend to).
- It's unlikely that the average person will have enough specific knowledge to be equipped to judge the truth of many scientific claims.

One implication of these two statements is that each of us can easily find ourselves in one of two positions: that of explaining something you know about to someone who is less of a specialist in that area than you are, or when it's you who is in the position of relative ignorance, having something explained to you. In these sorts of circumstances, the person with the knowledge is given our trust, as they could quite easily tell us things they know to be untrue—and we would be unable to separate truth from untruth in many of these cases. An immediate lesson here would be that it's vitally important to choose your authorities carefully, in order to maximize your chances of being told the truth. A later chapter (Chapter 6, "Why Can't We Trust Our World?") explores some of the reasons why this isn't always an easy task.

The second of the two statements raises a significant difficulty for someone who wants to be a responsible epistemic agent (epistemology is the philosophical field that addresses questions of knowledge—how and why we believe in propositions, and how best we should do so). As the boundaries of knowledge have expanded, theoretical fields have become increasingly specialized (Malone, Laubacher, & Johns, 2011). Instead of being an economist, you can now study to be a game-theorist; instead of a psychologist, a neuropsychologist; and the *generalist* in a field will find himself or herself increasingly ill-equipped to join into various conversations. Each of us has limited time and capacity for learning, so whatever we don't specialize in becomes something about which we are, to some extent at least, in a position of *vulnerability to false beliefs* with regard to that field of knowledge, thus needing to trust someone identified as an authority in that area. This need not entail anything sinister—we don't mean to argue that people are determined to deceive us (although sometimes they are), but more that we simply might not know the difference if they were.

The key insight here is that knowledge cannot be separated from power—having knowledge confers power, and lacking knowledge removes it and creates vulnerability. Furthermore, those with power (whether it be financial, moral, or political) can abuse it, sometimes endorsing knowledge and sometimes suppressing knowledge for their own reasons (Martin, 1999), such as personal gain. For those of us without such power, a clear implication is to be wary of believing something just because the person telling us to believe wears a scientist's coat, a priest's robe, or a presidential seal. There need to be good reasons for believing a claim, irrespective of who offers the claim. In logical terms, believing something because of who said it rather than what is being said is the fallacy of *appeal to improper authority*.

The good news is that even if one knows little or nothing about a particular field, there are still ways of determining whether a point of view is likely to be sensible or worth taking seriously, and talking about those ways is a fundamental motivation for our having written this book. We will never be able to avoid relying on people we believe to be authorities, but we can certainly learn how to choose our authorities responsibly—to become "authorities on authority," in a manner of speaking. But in doing so, one of the things we have to consider is that hyperspecialization and the deluge of information available to us might require that we shift our focus away from being *certain* that something is true. We'll seldom be in a position for such a strong judgment, and might instead want to adopt a less demanding standard, asking instead whether an idea or argument seems better *justified* than its competitors.

JUSTIFICATION

Any opinion worth holding—if you care about the truth, that is—must have some justification, which can be broadly understood as support or foundation for an opinion. The extent to which justification is present can be easy to agree on in fields that involve recognizable and measurable data (e.g., an engineer's claim that the bridge will not collapse), but often attracts disagreement in contexts such as moral argumentation, where "facts" seem thin on the ground.

It might be a mistake to think that we can't apply the same principles across these domains of knowledge, though. Claims made in domains such as morality or aesthetics are also capable of being less or more justified (Tramel, n.d.), just as claims in science are—even if we can't reach the same level of confidence regarding their truth.[1] What I mean is, although there may be *fewer* facts underpinning a moral judgement compared to one in science, a view that disregards what facts there are would still be inferior to one that takes those facts into account. We are, in other words, still responsible (if we want to hold responsible opinions) for digesting the information that there is as objectively as possible, and for evaluating the consequences of our views as clearly as we can.

The belief that we can hold more or less justified views regarding subjects like morality meets with plenty of opposition. Perhaps as a way to avoid these arguments entirely, you'll find that people often end up "agreeing to disagree," saying things like "everybody is entitled to an opinion." But in what sense is it true to say that "everybody is entitled to his or her opinion?" Or, to put it a different way, that "everyone has the *right* to an opinion?"

One possible way of interpreting these claims is that everyone, legally, is entitled to or has the right to hold whatever opinions he has. But that is surely not what we mean as, first, we'd have no way of knowing what opinions people held until they told us, and second, we'd only start caring once those opinions had some effect on real-world welfare.

To clarify what we mean, consider the occasions when certain phrases are used. We typically say: "Well, you're entitled to your opinion" when an opinion has been expressed and we *already* disagree with the expressed opinion. We then indicate that disagreement through expressing that other people have the right to disagree. If we agreed with the other person's opinion, we would just say so. Now, one way of framing this issue would be to point out that this conversation shortchanges our own opinions to an extent—instead of saying: "You're wrong. This is what you should believe," we find a way of saying something similar, but in a way that is less likely to cause offense. But for our own opinion to be worth holding, surely we should be willing to defend it more strongly than that. If we are that unpersuaded or committed to our own opinions, we should arguably not be saying "You have the right to your opinion," but rather "Oh, I hadn't thought about it that way—I'll reconsider my opinion in light of what you've said." Because if we're not willing to revisit our opinions in the light of new evidence, what are we doing holding those opinions in the first place?

Alternately, this "right to an opinion" response could be read as being dishonest, in that we might be saying something like: "You're just wrong about what you believe. And although I am right on this issue, I can't be bothered to argue about it." We say this can be dishonest because although you clearly don't respect the other person's view, you suggest that you do by saying that she has a right to her opinion. Even worse, this response could

---

[1] Some readers who are familiar with moral philosophy will disagree strongly with this, but I would plead their indulgence given that this book's primary topic is not moral philosophy, but critical thinking and psychology. For the purposes of this chapter, a coherentist approach to morality is assumed.

also be read as being demeaning toward the *person* you are speaking to, as you might be saying that she is simply too unreasonable to see sense on the issue in question. In all these permutations, we'd instead hope that we have strong justification for our own views, in that we have reasoned through the issue and settled on a well-considered position, rather than, for example, having believed something for so long (maybe because you heard it from a parent or a priest) that you are already completely committed to its truth, and argue to justify it rather than to assess its worth (Haidt, 2001).

Taking this a step further: Once we realize that what we mean by everybody "being entitled to his opinions" is so meaningless or unhelpful, why do we keep saying it? The immediate answer would be that it's a pleasantry—something we say to keep conversation going while avoiding a fight. But if everybody knows that we don't mean it in any meaningful sense, isn't its value as a tool for social harmony decreased?[2] If we all cared about holding well-justified opinions, we might instead say something like: "No, not everybody is equally entitled to an opinion. I've thought about my opinion for a long time, and can offer you reasons for my opinion being a better one than yours. So, in this case, when discussing this issue, I'm very happy to say that I am more entitled to holding an opinion than you are, or at the least that my opinion is the superior one."

RELATIVISM

So why don't we ever say this, or something similar? Partially, this is because it's become very unfashionable in the modern world to say that you know better than someone else. This is in some ways attributable to the global conversations we can now have thanks to the Internet, and also to the hyperspecialization discussed earlier. As we have become exposed to more and more different people, cultures, and religions, we have good reason to become less certain that our ways and our beliefs just happen to be the correct ones. Although this is a good thing in moderation, as it does encourage open-mindedness, the end of the line for that sort of reasoning is to say that no one is entitled to say that he knows the truth or has the best answer. In other words, it's a line of reasoning that ends in relativism, where "truth" is perceived as being relative to you and your particular situation—your history, your culture and upbringing, your influences. But there is no need to take this relativistic line, and there are, in fact, good reasons for not doing so (Boghossian, 2006).

---

[2] There are numerous examples of this in the English language, where what is said is actually the opposite of what is meant. One of our personal favorites is the phrase (often used in the American South): "Well, bless her heart!" On the surface it seems nice and lovely, but in reality is almost exclusively used to avoid saying someone's moronic.

There is no need to endorse this line of thought, because even though it may be difficult to agree on the truth of the matter, this does not mean that one opinion cannot be inferior to another, or that there are some opinions that just should not be taken seriously. And the most obvious good reasons for not taking the relativistic line are that treating claims relativistically removes part of our opportunity to correct our own errors through learning from others, and to discover when we are being complacent about our beliefs or reasoning. Relativism also takes away our opportunities to inspire other people to be responsible about what they believe. But why would something like personal belief in something matter, and why should we care about what others believe?

IF ACTIONS MATTER, BELIEFS DO TOO

It is vitally important to care about what we believe. For all the emphasis that is placed on action, we can forget that action is (usually) motivated by thought: We desire, we need, we plan, and then we act. If we don't think about what we believe, the chances of arriving at optimal actions are impacted. Our self-interest motivates us toward thinking clearly about what's best for us personally, and selecting the actions that are best suited to achieving the outcomes we desire. In other words, what we believe does matter; as does what other people believe, because those beliefs typically translate into real-world actions. Although we can tell people that they are entitled to their beliefs, we should remember that this is mostly a political claim, in the sense that we are affirming human equality and committing to respecting others through not prejudging their opinions to be worthless. We are not, and never were, saying something about how all opinions are literally equal in worth, even after we've been exposed to their respective merits.

Consider an example: Fred and Bob both claim that the president of the United States is going to lead the world into a nuclear war. Fred says he believes this because he has read statements from U.S. military officials that indicate America is stocking up on missiles, because of psychological reports he has read about the character of the president, and so forth. Bob says he believes this because last night, when he was asleep, a wise gray squirrel appeared to him in a vision and told him that the president would start a nuclear war.

When thinking about that example, it would be easy to get sidetracked by your current knowledge or beliefs about an existing president and his or her character, as well as your beliefs about the current state of international politics. Note, however, that being sidetracked by these factors would be a mistake. The thing that makes Fred worth talking to (and that makes Bob worth ignoring, at least in terms of the truth of his statement) is that we can understand the methodology behind the forming of his opinion. Fred has taken evidence, thought about that evidence, and come to a conclusion. He may have done so badly, or somewhat irresponsibly, but he is still speaking in a language we understand—the same language we use when taking claims seriously in science, whether it be economics and psychology (social sciences) or biology and physics (physical sciences).

We know that there would be a point in debating with Fred: We can dispute his evidence (the psychological reports, for example) or his reasoning (the way in which he uses the evidence to reach a conclusion). Debating with Bob would merely involve telling him that the gray squirrel (if it exists) has no good reason to believe in the likelihood of certain events, and that Bob has no good reason to believe what the squirrel says. (From what we know of Bob, he won't much care about these technical details.) So, in the end, it doesn't matter one bit whether Bob and Fred are right or wrong—they could both be right, or both wrong, but although we can learn something from talking to Fred (and he from us), it would most likely be an inefficient use of our time to talk to Bob (at least assuming our impetus is our care for the truth, rather than our wish for entertainment).

OPINIONS, BELIEFS, AND KNOWLEDGE

We are constantly forced to make choices. Fortunately, many of these choices have no serious negative consequences for us, given that they are typically directed at our best interests. Our decisions can and do affect others also, of course—not only directly with regard to their welfare, but also in shaping their perceptions and responses to us. We'd like others to treat us well, and we'd like to make the best decisions for ourselves (and others), which means that it is crucial to be as well informed as possible when making decisions. This means not only having as many of the facts at your disposal as possible, but also to treat those facts responsibly, rather than conveniently ignoring those facts that you are uncomfortable with, or would prefer not to believe.

But what is it to be informed? One way of answering this question is to say that you've achieved a sort of synchronization between the contents of your head and the world outside—that you (largely) believe those things that are true and disbelieve those things that are false. Unfortunately, we sometimes get this order wrong. We sometimes respond to the world as if it should correspond to our beliefs, rather than treat our beliefs as if they need to correspond to the world. This reluctance to change our minds in light of evidence has been described as so pervasive as to suggest that we're living in an "age of willful ignorance" (McIntyre, 2015). However, you don't need to be that pessimistic to nevertheless agree that we are too often reluctant to change our minds in light of evidence that our beliefs are false. We prefer to think that the information we are getting is false, rather than thinking that something we believe may be false, and that is why the world doesn't act the way we think it should. This habit is an example of *confirmation bias*—the tendency to easily view evidence as confirming preexisting beliefs and to be reluctant to understand evidence as disproving beliefs we currently hold.

Why would we be so lazy or complacent in choosing what to believe? Often the answer seems to relate to the fact that it often doesn't much matter what you believe in practical terms— other people are usually willing to continue to talk to you or do business with you, whether or not you believe something different from them. Another part of the answer may be that many of our beliefs are never exposed or tested. We are not given an opportunity to engage with or question them, because they simply never come up, or if they do, they come to the surface in contexts in which it's easy to not bother challenging them. But whether or not it's easy to understand *why* we often simply carry on believing the things that we currently believe, the more important issue is whether we *should* be so lazy or complacent in choosing what to believe.

One way of approaching this more important issue is to consider how we feel about different sorts of beliefs. Take, for example, an engineer. It is surely the case that we care about many of her beliefs. If she believes that her design for a particular bridge is sound—in other words that it can support your car as you travel 65 feet above another highway—we have some quite serious reasons for being concerned that her beliefs do in fact map onto the world and its physical laws accurately. If your employer believes that you are lazy and incompetent, we likewise care about the accuracy of that belief. For some reason, though, we don't seem to care in the same sort of way about other beliefs, particularly beliefs relating to morality, politics, aesthetics, and metaphysics. They matter, in that we may choose to interact with people who happen to share the same sorts of beliefs that we do in these areas (e.g., supporters of a particular football team might prefer to frequent a bar where other supporters of that team congregate), but for the most part, these disputes don't seem to affect many of the day-to-day events of our lives, and they also aren't interrogated as strongly regarding their correspondence with the real world. Everyone is entitled to his or her opinions, right?

So what is the difference between these different sorts of beliefs? Some may think that the answer was contained in the observation that much of the time, aesthetic beliefs (e.g., your belief in your team having the most dynamic players in the league) simply don't affect our lives in the way that the construction of a bridge does. But is this true? Isn't the difference between these two sorts of beliefs more that we tend to treat the engineer's beliefs as *facts* and your beliefs about football as *opinions*? And then, surely, some opinions are actually *true* in the same sort of way as the engineer's beliefs are, and other opinions are simply *false* or indeterminate—so not only can we make distinctions regarding the quality of opinions, but this also suggests that some opinions are surely far closer to being "fact" than others.

The point here is that we tend to use words like "belief," "opinion," "fact," "true," and "knowledge" in very casual ways (McBrayer, 2015), which allow for you to be talking about something different to me when we both use words like "opinion." One of the points made earlier is that we minimize our chances of making mistakes if we fill our heads with pieces of information that are actually *true*, whether those pieces of information are called beliefs, opinions, or facts. Even if you end up disagreeing on any given taxonomy of those words, a concern for justification should be central to your understanding of all of them. The reason for this is that all of our beliefs are connected to other beliefs, and most of them are (sometimes in very roundabout ways) connected to some very foundational beliefs that we hold, beliefs about very essential or basic things. For example, take your belief that your name is "Bill." Your belief that your name actually is Bill (or whatever your name is) depends on various other beliefs, such as that your "parents" are actually your parents; that if they are your parents, that they are trustworthy; and so on.

The point in questioning your belief in something as basic as your name is to make it clear that not only are beliefs interconnected, but also that a mistake in the foundation of your belief network can corrupt any beliefs that depend on that initial belief. If this is true, it means that we *should* care about all of our beliefs—or more specifically, about whether they are true or not. And by "true," we can't mean something like "true for me, but maybe not true for you," because that's not what the word "true" *means*. When we say something is true for me, but maybe not for you, we are not actually talking about "truth" as commonly understood, but rather saying that I *believe* it, and perhaps you don't.

These issues wouldn't matter if our beliefs made no difference to the ways in which we interact with the world. But history offers us countless examples of ways in which beliefs—particularly about things with little or no empirical evidence to support them—can have serious and sometimes catastrophic effects on the world. For example, consider the belief, held by many even today, that one particular race or gender is superior to another. Or consider the belief that the measles, mumps, and rubella (MMR) vaccine causes autism: even though it's based only on poor and discredited science, along with some misguided celebrity endorsement, it can—and has—caused deaths from a disease we thought was already defeated in the United States (measles).

Yet, despite the effects they can lead to in the world, we don't treat beliefs symmetrically (in that some are interrogated more critically than others), and we also offer beliefs exemptions from standards we'd hold "facts" to—you can believe whatever you like, so long as you don't perform actions that harm other people on the basis of those beliefs. The broader point being made here is that if we are willing to offer some beliefs exemptions from corresponding to evidence, should we not extend the same charity to *all* beliefs that can't be proven to correspond to the world? And, if not, how do we choose which are exempted and which are not? A principled way to distinguish between those two categories is important, because one person's demand that she be entitled to believe in a pseudoscience is otherwise difficult to distinguish from the racist's demand that he be regarded as superior to you. Either these cases can both be assessed on their internal logic, or they need to both be assessed with reference to some *external* standard (possibilities here include both correspondence to the external world and capacity to result in benefit or harm).

The general point is that a very simple principle—consistency—demands that we treat things that are similar in a similar fashion (Ryan, 1996). The challenge of argument here is that we'd either need to show that these examples are not similar or, if that fails, to accept that they need to be treated symmetrically. If you don't subscribe to either of these beliefs (in a particular pseudoscience or in scientific racism), it's no doubt far easier to see their similarity in both being unjustified by external evidence (rather than their own internal logic). They are also similar in that pseudoscientific beliefs can affect other people as negatively as racist beliefs. And if we are then to accept that these beliefs are similar in these sorts of ways, consistency demands that we treat them similarly by either respecting *all* relevantly similar beliefs that don't correspond to evidence (or where evidence is difficult to find), or that we are skeptical about all of these sorts of beliefs.

Being skeptical of such beliefs does not necessarily mean believing that they are false. It simply means not being dogmatic about them being true, and being open to the *possibility* that they are false. Of all such beliefs, it may well be that some (or even many) of them are in fact true, but the problem remains that we cannot prove them to be so. This means that if you are going to believe them, you commit yourself to a certain amount of risk, especially if you are going to perform actions motivated by these beliefs. Again, though, it needs to be made clear that if one is prepared to believe things on the basis of little or no evidence, one increases one's chances of believing things that are false. This is hardly a good strategy for being informed, and thus increasing your chances of making optimal decisions.

Back to our engineer: a key feature of her claim, and one that makes it "knowledge," is that it is true—or that it is so likely to be true that we may as well call it true (certainty being too demanding a standard). What prevents my (JR) claim that the death penalty is unjust from being knowledge is that I cannot demonstrate it to be true to the same extent. Of course I can argue for that belief, and I may well be able to persuade you that it is a good belief to hold, but I will not be able to *prove* it to be true beyond any reasonable doubt. So it remains in the realm of opinion, rather than fact, although it's certainly the sort of opinion that is worth taking more seriously than the opinion that, for example, men are (literally) from Mars and women are (literally) from Venus. Opinions can vary in quality, and can be treated with the relative respect that quality earns. In this case, my opinion would be supported by evidence, good reasoning, and the like—I can argue for it, and offer you reasons why it's a good belief to hold. So a large part of our answer as to which opinions to take seriously has to do with how responsive the opinions in question are to evidence, in other words, how well *justified* they are.

So far we have considered four elements of what one could call our mental furniture: opinions, beliefs, knowledge, and justification. Let's now try to define these more carefully in an attempt to understand them in relation to each other.

Opinions and Beliefs

There are certain properties we would like our opinions to have—we would like them to be true rather than false, and we would also like for them to be based on good reasons rather than bad reasons. Opinions are also propositions or statements that we believe to be true—if I say "the death penalty is unjust," or "adulterers will go to hell" I am expressing an opinion and at the same time expressing a belief. After all, something cannot be your opinion if you do not believe it to be true. So for our purposes, a belief can be treated as meaning the same thing as an opinion; some of our opinions (or beliefs) could be true and others could be false, and we could also hold varying degrees of commitment to our own beliefs (being more convinced of some than of others).

A belief based on superstition, for example, is based on poor reasons unless that superstition happens to be true. Even if that is the case, the belief is still based on poor *reasoning*, in that even though the superstition ends up being grounded in fact, you're still reasoning poorly if you appeal to the superstition itself rather than those facts. Believing something based on poor or no reasons could also be said to be *irrational*, which often boils down to simply claiming more than is suggested by the evidence available. And, if you form opinions in an irrational fashion, you increase your chances of having false opinions—you're in effect buying lottery tickets rather than weighing the evidence. As discussed earlier, given that our beliefs and opinions affect not only how we form subsequent beliefs, but also how we *act*, we do need to care about holding true beliefs rather than false ones. For those who like definitions, let's summarize in saying that a belief consists in having a clear disposition toward the truth value of a statement.

Knowledge

Some of our beliefs are, in fact, true and others are false, regardless of our views on the matter. We (in everyday life) tend to regard a belief as true when it is consistently demonstrated to be true (e.g., the law of gravity: I believe this to be true because things I drop consistently fall to the ground), and are more cautious of believing claims to be true when it is difficult to demonstrate their truth. But, regardless of our particular justification for regarding a claim as knowledge, it would be distinctly odd to allow for claims that are *false* to count as knowledge. We might (mistakenly) think they are true, and regard them as knowledge, but we'd simply be wrong about this.

As with many of these concepts and debates, we should be careful to separate how we talk about the concepts from how they are most usefully understood—"knowledge" is most usefully reserved for things that are factual and known, rather than strongly held beliefs that are possibly false. It is maybe simplest (for everyday purposes) to say that a proposition counts as knowledge if (a) you believe it for good reasons, and (b) it is actually true. The philosophical literature on the topic and the complexities raised therein are far beyond the scope of this text, as you can verify by reading an elegant and oft-cited paper by Gettier (1963), interrogating a similar conception of knowledge as "justified true belief."

Following Gettier, it is important to note that for something to count as knowledge, you must believe it for good reasons. If, for example, I have a dream in which the gray squirrel says to me that I will win the lottery, and then I do win the lottery, it was never the case that I *knew* that I would win the lottery. I believed it, and perhaps even strongly believed it if I place great faith in the gray squirrel, but I never *knew* it—not if by "know" we mean something more serious such as, "I know that tomorrow is Friday" (if today is Thursday, of course!). So even though we may often use the words "I knew it" in casual conversation, we are using the words "know" or "knowledge" in a casual fashion, and "know" means something quite different when we say "I know that water boils at 100 degrees Celsius" (at sea level and so forth). Again, the difference between these two sorts of claims is that my belief that I would win the lottery was *poorly justified*, whereas my belief that the following day would be Friday was *strongly justified*.

The second element of knowledge previously mentioned is truth—for something to count as knowledge you need to have justified belief in something that is *also* true. Here, we can simply say that a statement is true if it actually corresponds with the way the world is. Note that corresponding with the way you *think* the world is, or the way that you would *like* the world to be, does not make something you merely believe count as knowledge. So we need to be careful to avoid subjectivity here, and remember that if a proposition corresponded with the way the world actually was, it would probably be the case that most everyone agreed with it (the laws of gravity are again a good example), rather than that it was something nobody else but you believed to be true. What makes a proposition true is that the world actually looks or acts like that—no matter who happens to believe what.

This raises a serious philosophical issue though: It is certainly possible that we don't have access to certain facts about the world that we would need to really know when a belief is true or when it is false, and thus, when it can be regarded as knowledge. To return to the example of religious beliefs, even if we can't prove that the world actually corresponds to those beliefs, it might still be the case that those beliefs are true. Or to take another example, it might possibly (in a logical sense) be the case that it actually *is* unlucky to walk under a ladder. Perhaps there is some law of nature that dictates this unlikely principle, even though we would struggle to find evidence for it.

Regardless of those complications, if there is no evidence for a belief, it is *irrational* to believe it—*even though it might actually be true.* Likewise, we can make mistakes of the other sort—we can believe something because it seems to be justified by the evidence, but then we eventually learn that the belief in question was false. If you were living before a time when we had boats that could travel long distances, and we had no telescopes or photographs from space, all the evidence you had access to might have led you to believe that the Earth was flat, even though that belief turns out to be false. By contrast, to believe that the Earth is flat today, given the evidence available, would be an unjustified and irrational belief.

Justification of Beliefs

Justification relates to the evidence we have for holding a belief. Here, it is important to note that different sorts of evidence may be required or appropriate to different sorts of belief. If you are trying to persuade me that the unemployment rate is likely to decrease by 10% over the next 3 years, the evidence you use is likely to be very different from the sort of evidence you use to persuade me that there is life on other planets. Regardless of this, either or both of these beliefs can be well justified (or either or both can be poorly justified). A second complication is that although there may be information that helps us determine whether a belief is well or poorly justified, we may sometimes not have access to (or not understand) that information.

This raises two issues. First is that our beliefs can only be justified to the extent of the information available to us. Second, that if we are to claim that something is true, it is our responsibility to accumulate enough evidence related to that belief to entitle us to claim that the proposition in question is true. If you have not bothered to do the homework, you are certainly still entitled to claim something as an opinion or belief, but you are by no means entitled to claim that it is true, or that anyone else should believe it also.

The more important lesson here is perhaps for us to realize that before we regard something as true, and especially before we form other beliefs (or are motivated to action) on the basis of a claim, that we do the necessary homework in justifying our belief in that claim. Here confirmation bias, discussed earlier, is again relevant. We tend to treat our existing beliefs as already justified, forgetting that the mere fact that we believe them does not in itself count as justification. We need to think about why we believe what we do—if we have good reasons, which are likely to be accepted by other rational people, we are justified in believing something. If we don't have such reasons, we should acknowledge that our belief is unjustified, and not insist that other people take that belief seriously.

WHAT'S THE POINT OF CRITICAL THINKING?

Some people may want to argue that it's not important to think about thinking, or to pay attention to issues such as the differences between beliefs and knowledge, and the role justification plays in distinguishing them. One could be tempted to think, for example, that this sort of reflection doesn't get any of the world's work done, and is instead a leisure activity for those who have nothing better to do. Fortunately, there are three easy answers to this sort of skepticism, which we'll adapt from Blackburn's *Think* (1999).

The High-Ground Answer

There are many things we do in life that don't directly serve some pragmatic purpose. We might listen to music, watch films, or play sports—and in these cases, it's exceedingly rare to find someone asking what the point is. There are different sorts of possible "points," and it's not necessary for everything to relate to the global economy for it to be relevant to human life. Our lives consist of various elements, and one could argue that the activity of thinking, and of trying to understand human nature, is interesting and rewarding for its own sake. Blackburn also suggests an analogy comparing mental exercise with physical exercise: just as physical exercise makes the body fitter and stronger, causing us to feel better, so mental exercise can cause us to feel better by making the mind more stronger and more flexible.

The Middle-Ground Answer

What we think—the contents of our heads—is a key contributing factor for determining what we do. As discussed previously, our beliefs with regard to political or religious questions can have serious and sometimes catastrophic implications in terms of our lives and the lives of others. It's undeniable that we care about how people act—and given this concern, it follows that we need to care about how people think. This is especially true for ourselves, seeing as we have a direct investment in making decisions that are in our own best interest.

The Low-Ground Answer

The low-ground answer takes the sentiment of the middle-ground answer, but asks you to reflect on the fact that people can be persuaded to do quite nasty things to each other, often as a result of being encouraged *not* to think, but just to believe. Plus, there are so many more other people in the world than there are "you's" in the world—so you (and I) would far prefer that they pay attention to reason and evidence. Critical thinking is our best antidote to the sleep of reason, and is therefore clearly worth cultivating.

CONCLUSIONS

With the definitions given in this chapter, and especially if you agree that it's difficult and often impossible to possess absolute certainty with regard to any claim, this discussion points to justification as being the most important element in terms of our being able to say that we "know" something. "Truth" becomes less relevant—not because it's unimportant, but because it's more difficult for us to establish whether something *actually is* true or not than whether we have compelling reasons to regard it as true.

As soon as justification takes center stage, it becomes easier to recognize the similarities among different forms of knowledge, and how it is possible for the engineer to say "I know this bridge can support x tons of mass" and at the same time for you to say "I know my name is Bill," or even "I know that the death penalty is immoral." In each of these cases, there is information available that entitles one to claim knowledge, so long as she has responsibly gathered sufficient evidence and, as objectively as possible, considered whether that evidence, presented in a cogent argument, adds up to a justified claim.

Of course, it may well be that for some issues (tricky moral questions like human cloning, or the existential threat of artificial intelligence) the evidence is too difficult to find or understand. But that does not mean we can't arrive at better and worse justified claims in these areas—it just means that we have to work far harder to derive them. We should also not forget that just because we cannot bring ourselves to consider something "knowledge" right now does not mean that it will not eventually be regarded as knowledge. Technology advances, we discover new data, and our arguments improve. In the past, we were unable to prove that there was no material reason to discriminate unfairly based on race and gender, and now we can.

Examples such as this should remind us that we are learning things all the time, and that we should remember that there is still much to learn. Even if we can't know (in other words, demonstrate it to be a justified belief) now that, for example, it is possible for humans to live on other planets, we are nevertheless in the process of gathering information that may make it possible for us to know this 100 years from now. That continual search for knowledge—and the revision of false beliefs in light of new evidence—is the essence of science and the scientific method, which we'll discuss in the next chapter.

QUESTIONS FOR REFLECTION

1. *Are false beliefs always best avoided or eliminated, or can you think of exceptions to this principle?*
2. *Are there any principled grounds by which false beliefs should be exempted from the general ideal that we should aspire to only hold true beliefs?*
3. *Do you think that humans in the 21st century are more susceptible to holding false beliefs than they were in the past? If so, how do you think we might best respond to modern challenges related to conflicting information and information overload?*
4. *Consider different sorts of belief: for example, belief in a personal god (or gods) versus belief that smoking contributes to cancer. Are these beliefs justifiable (or not) in the same sorts of ways, or are they different kinds of belief, with different standards of assessment?*
5. *Is there anything that you regard as being certain knowledge—in other words, something about which you allow no possibility that you are wrong? If so, how do you justify this certainty?*

REFERENCES

Blackburn, S. (1999). *Think: A compelling introduction to philosophy*. Oxford, UK: Oxford University Press.

Boghossian, P. (2006). *Fear of knowledge: Against relativism and constructivism*. Oxford, UK: Oxford University Press.

Gettier, E. (1963). Is justified true belief knowledge? *Analysis*, 121–123.

Haidt, J. (2001). The emotional dog and its rational tail: A social intuitionist approach to moral judgment. *Psychological Review, 108*(4), 814–834.

Hampton, K., Rainie, L., Lu, W., Dwyer, M., Shin, I., & Purcell, K. (2014). *Social media and the "spiral of silence."* Retrieved from http://www.pewinternet.org/2014/08/26/social-media-and-the-spiral-of-silence

Malone, T., Laubacher, R., & Johns, T. (2011). The big idea: The age of hyperspecialization. *Harvard Business Review, July–August*. Retrieved from https://hbr.org/2011/07/the-big-idea-the-age-of-hyperspecialization/ar/1

Martin, B. (1999). Suppression of dissent in science. In W. R. Freudenberg & T. I. K. Youn (Eds.), *Research in social problems and public policy* (Vol. 7, pp. 105–135). Stamford, CT: JAI Press.

McBrayer, J. (2015, March 2). Why our children don't think there are moral facts. *The New York Times*. Retrieved from http://opinionator.blogs.nytimes.com/2015/03/02/why-our-children-dont-think-there-are-moral-facts

McIntyre, L. (2015, June 8). The attack on truth. *The Chronicle of Higher Education*. Retrieved from http://m.chronicle.com/article/The-Attack-on-Truth/230631

Ryan, S. (1996). The epistemic virtues of consistency. *Synthese*, *109*(2), 121–141.

Tramel, P. (n.d.). Moral epistemology. *Internet Encyclopedia of Philosophy*. Retrieved from http://www.iep.utm.edu/.

Trout, J. (2002). Scientific explanation and the sense of understanding. *Philosophy of Science*, *69*(2), 212–233.

CHAPTER 2

WHAT IS SCIENCE?

HYPE VERSUS HYPOTHESES

We're bombarded with claims made in the language of science every day of our lives. Newspapers tell us what's safe or unsafe to eat, popular magazines routinely feature colorful pictures of our brains, all in an effort to explain—or gesture at explanations of—claims regarding things like what makes us happy or what we're addicted to (or could become addicted to, if we're not careful). But a general problem with these media representations of scientific activity is that they often make things appear much simpler than they actually are. Science doesn't advance one press release at a time, despite appearances to the contrary (Goldacre, 2014)—you might never hear of significant breakthroughs, and what you hear about might in turn be misrepresented or deeply contested. Despite these complexities, there are some simple principles underlying scientific endeavors, as outlined in this chapter. However, the actual work of doing the science can be slow and uncertain, and the conclusions reached can be far more tentative than the headlines might lead you to believe.

Perhaps most important, the public typically won't get to hear of all the uncertainties and qualifications generated by scientific inquiry, and it is unlikely that the public will be aware of the fact that a significant proportion of published research findings—perhaps even the majority of them—are in fact false (Ioannidis, 2005). This is primarily because these complexities don't provide as compelling a narrative. Who wants to read about how "scientists are still uncertain about the strength of the causal connection between total cholesterol in the blood and heart disease," versus "eat as much fat as you like! New study shows that cholesterol is not a risk factor in heart disease!" The more hyperbolic and dogmatic a claim, the more it seems to dictate what appears on magazine covers and in dinner-table conversations. But this phenomenon, which we might describe as a "sexying up" (via simplification), in which titillation is what matters most, presents the danger of having us mistake the controversial or particularly exciting cases as being representative of how science *typically* works, and of the scientific method in general. The reality is that our knowledge of the world is always contingent on what we know—and can know—at any given time. By contrast, popular media debates on scientific matters can make it appear that we're constantly replacing one dogmatic conclusion with another, in light of some new piece of data or study.

Instead, we need to remember that it's more accurate to regard new evidence as *tipping the scales* one way or the other, with a conclusion becoming more or less likely as new evidence emerges. It's usually a mistake to speak of being certain of some conclusion, both because certainty is very rarely (if ever) available to us, and second because it encourages an attitude of dogmatism, rather than the skeptical and curious attitude that is the hallmark of good scientific practice (Popper, 2002). Of course some conclusions are so well established that they might as well be regarded as certain, but the fact remains that contrary evidence will still tip those scales, even if to a trivial degree. Openness to correction is a key virtue that distinguishes science from pseudoscience (see Chapter 3 for more on this), so it's important that our language reflects that we are able to change our minds.

None of the preceding is an objection to making science compelling and interesting, even sexy and cool, because it is! Part of the purpose of this book is to inspire people, especially young people, to get involved in scientific work, as well as to help everyone understand scientific conclusions and live evidence-driven lives. But making science "cool" should not come at the cost of dumbing it down, or making it appear simpler than it is. We shouldn't be encouraged to think that getting clear positive results from a study is the norm, for example. Much scientific research will achieve few concrete results in terms of telling us something new or confirming a counterintuitive hypothesis, but could nevertheless help us to rule out various other hypotheses as *not* worth pursuing. This would still be a significant contribution to the body of knowledge, in that it would allow for future researchers to use their time more productively rather than chasing what have now been established as dead ends. But although the identification and elimination of these dead ends does constitute "new" knowledge, it is knowledge that will be of more use to future scientists than it would be for the lay public, and might not achieve the same prominence in the public media.

The general point is that much of science involves a slow, long slog and is conducted by people outside of the limelight, who often shun the limelight because they care more about the work than about becoming celebrities. The difficulties and challenges inherent in some fields of inquiry is exactly what motivates some scientists to work so hard to find the answers to questions posed in those fields. As examples like Carl Sagan or Neil deGrasse Tyson show, there's certainly no necessary tension between celebrity and science (Fahy, 2015), it's just that they tend to not be strongly associated with each other, thanks to the complexity of the work involved.

But even in cases in which the work is difficult—even perhaps completely obscure to nonscientists—there are nevertheless some general principles underlying scientific inquiry that we can all benefit from understanding. This chapter seeks to summarize those principles, with the aim of facilitating a broad understanding of why some apparently scientific claims, and people purporting to be scientists, are worthy of your attention and trust, and others less so.

SCIENCE: NOT ALWAYS "COMMON SENSE"

Our beliefs and attitudes are informed by the experiences we have, but it is easy to forget that those experiences might not be representative of what's typical for humans more generally. What this means is that a very small set of data (we are only one data point among billions) can be used to develop conclusions that perhaps don't accurately reflect the beliefs and attitudes of others. Now, of course it's true that our circumstances might be atypical, and therefore it's often true (and entirely reasonable) that our experiences and expectations *should* be different from the norm, or from those of other people we encounter.

But precisely because of how powerful our own experiences are to us—being the only ones we have direct access to—we can also be led to overstate the value of personal experience and forget that we *shouldn't* expect the patterns of our lives to be typical for others—or even reliable when it comes to understanding our own lives (Mlodinow, 2009). To take a simple example: If you have great success in losing weight while eating or not eating some sort of food (perhaps bread), you might think this would be true for everyone. This conclusion would be unduly hasty, for at least two clear reasons.

First, although human biology is of course similar, there is nevertheless enough variation among us that generalizing from your (single) case to a population of billions is exceedingly risky. Second, your anecdotal experience isn't usually as reliable as you might think it is. As mentioned in Chapter 1, our tendency is to overemphasize evidence that confirms something we already believe to be the case, while underemphasizing contradictory evidence: We'd typically be more likely to give credit to the diet while perhaps forgetting (or underplaying) the extent to which the dieter also simply ate less, or exercised more, or just contracted a tapeworm. What we perceive as the most *proximate* (closest in relationship to what we're experiencing; or immediately apparent as related) cause of something we experience—in this case weight loss—might be obscuring our recognition that something else better explains that experience.

The general point is that we tell ourselves stories to make sense of our lives and the world. And this general point reveals a key difference between how science operates (or should operate) and how common sense operates.

In science, the evidence is the evidence, and leads us to a particular conclusion (even if that conclusion is "we don't know"). Common sense, in contrast, tends to lead us away from uncertainty, and toward beliefs that fit in with other things we already believe to be true. We're programmed to fit information into structures and patterns (Shermer, 2008), and although doing so can often be useful (consider medical diagnoses as example, where understanding symptoms as representing some broader condition can help the physician to identify what ails you), this pattern-seeking behavior (or *patternicity*) can also lead us to errors such as the "gambler's fallacy," in which we might keep laying money onto the poker table because we think we see some fortuitous pattern in the cards that have been dealt in previous hands.

A scientific outlook is not personally invested in certain outcomes or subjective perceptions of data—it will only take your anecdotal experience for what it's worth, and what it's worth is not much, given that it's merely one data point among billions. It's also not a reliable data point, given that the observations themselves can't be externally validated for accuracy. In the language of science, our (perfectly natural) habit of deriving conclusions from personal experience amounts to an *uncontrolled* experiment, in which the variables involved are difficult to impossible to identify or quantify (Chabris, 2011).

Think further about some of the rules of thumb that inform our lives, many of which identify principles or strategies for decision making that could be described as "common sense" (Lilienfeld, Lynn, Namy, & Woolf, 2013). To express the idea of being prudent, we have "Look before you leap." But we also have "He who hesitates is lost." The contradiction between these is clear, and leaving aside the fact that some people are naturally more risk averse than others (and will therefore tend to look before they leap), they lead us to recognize that these general rules always need interpretation and contextually aware application—simply applying a rule without regard for context is unlikely to result in an optimal outcome. There are dozens of similar examples, like "You're never too old to learn" versus "You can't teach an old dog new tricks," or "Birds of a feather flock together" versus "Opposites attract." Part of what the scientific outlook gives us is a way to help us navigate confusions or contradictions like these in a principled and fairly reliable manner by giving us the tools to explain and predict the relationship between causes and effects.

We say "fairly reliable" because we typically cannot guarantee the truth of a conclusion, the best we can do is to reach conclusions as responsibly as possible, and accept that following a sound methodology vastly increases the likelihood of those conclusions being true. It's no discredit to science that it gets things wrong on occasion—the point is, there's no *better* way to reach conclusions, and the scientific method gets things right more often than it gets things wrong.

WHAT IS SCIENCE?

This simple question has a relatively complex answer. An entire subfield of philosophy, the philosophy of science, is in fact dedicated to trying to answer that question. This text aims at a general explanation of issues in critical reasoning and scientific inquiry, so in contrast to specialist texts in the philosophy of science, we will not be assessing the relative merits of realism, antirealism, empiricism, rationalism, and the like.[1] Suffice it to say that your authors adopt a viewpoint that is broadly *realist* or *naturalist*, in that we believe that scientific theories typically should—and in fact often do—represent what is actually real. Broadly, you can contrast this perspective with antirealism, which holds that scientific theories are not dependent on, or obliged to, represent metaphysical reality, and that they can work to describe observations regarding things like electrons or black holes without necessarily being accurate.

We also adopt an *empirical* stance, which holds that observations and the generalizations we make on the basis of those observations are central to scientific inquiry. Because scientific theories are empirical, this means that they are responsive to new observations, even when those observations conflict with existing hypotheses or theories (terms that we'll define in a moment). One cannot dogmatically hold on to theories, especially when they do not accommodate or explain new observations. As will be explained later, this means that theories are always open to being *falsified*.

Another way of putting this point is to highlight the importance of *fallibilism*, which is the knowledge, as a scientist, that new data might still come to light which puts all of your current convictions into doubt. We therefore embrace the possibility of error, and reject a dogmatic attitude toward what we know or think we know (which, as we saw in Chapter 1, may or may not be justified). In fact, the words "know" or "knowledge" are more accurately understood as referring to things that are true to the best of our (current) knowledge, while acknowledging that we might have to change our minds about what is true at some future point. The notion of fallibilism reminds scientists that certainty isn't available to us, and also it reminds us that it's *good to be wrong*, because "nothing obstructs access to the truth like a belief in absolute truthfulness" (Deutsch, 2013). Given those preliminaries, how might we define *science*? A definition that we find particularly useful is:

> A set of methods designed to describe and interpret observed or inferred phenomena, past or present, and aimed at building a testable body of knowledge open to rejection or confirmation (Shermer, 2002, p. 145).

---

[1] Those interested in an overview of these distinctions and debates could usefully consult a text such as Okasha's *Philosophy of Science: A Very Short Introduction* (2002).

Shermer's definition captures the essence of what science or the scientific method entails. Science offers us a toolkit, and it's then up to us to use those tools correctly and responsibly. Another feature of this definition worth highlighting is the fact that scientific hypotheses need to be *testable*. As we'll see later in this chapter, if there's no way to test a claim (even if the manner of testing is currently hypothetical), that claim currently falls outside the scope of science, and always might.

This principle of testability is vital to the health of science, in that it allows for science to be self-correcting. We can only discover our mistakes if there's a way of discovering them, and that means that claims must be testable. On a personal level, thinking scientifically is useful for a similar reason: We're very good at fooling ourselves, and one way to minimize fooling yourself is to test what you believe against the facts. And, if there's no way of testing one or more of your beliefs, then those beliefs might be false, and you would have no way of knowing.

A concern for being able to test hypotheses points to a related issue, namely, *bias in methodology*. Given that we know, in advance of even attempting to test a hypothesis, that we are prone to confirmation and other biases (as discussed extensively in Chapter 5), it is incumbent on us to design our experiments and other scientific activities in ways that attempt to ensure that we correct for these biases. The fact that we know how important it is that we allow ourselves to be wrong doesn't itself offer any guarantees that in the moment, we'll happily accept error. It's not comfortable to discover that we hold false beliefs, and this makes it vital that our reasoning, and our experiments, are not set up in ways that minimize the chances we'll be proved wrong.

Good science—and good scientists—tries to exemplify the virtue of dispassionate truth seeking. This does not mean the scientist does not care about his work, but something else entirely. Being a good scientist requires integrity, in the form of being honest and scrupulous with data—by not misrepresenting what it says, and by not ignoring data that is inconvenient to one's hypothesis (Goldacre, 2012). Good science requires a kind of disinterestedness, in that you would try your best to not be influenced by personal commitments or financial interests, such as pressures from those who are funding your research.

Most important, perhaps, is that good science requires that we're willing to be wrong and even that we seek out opportunities to discover where we are wrong. In doing so, beliefs that are tested and *survive* that testing can be regarded as more reliable than they were before surviving the test. Because we're so good at fooling ourselves, this also requires that we collaborate with each other as scientists—other people might spot issues that you are unable to see, perhaps because of bias, or perhaps because you've just spent too much time looking at things in one way, and a fresh pair of eyes can see them in a different way. Science, in other words, is usually collaborative or communal (Adams, 2013). As a scientist, the point of publishing scientific findings in peer-reviewed journals is that the community can benefit from your findings, but also so that they can help you test them, and discover whether they can be replicated or not.

BUILDING BLOCKS OF THE SCIENTIFIC METHOD

Words like "hypothesis," "law," "theory," and "fact" are encountered fairly frequently in everyday conversation, even by nonscientists. But, these usages might sometimes be misleading or idiosyncratic when compared to their technical definitions. As such, let's start by briefly defining them as used when speaking about the scientific method.

A *hypothesis* is a testable statement that accounts for a set of observations. The task of testing it is made easier by stating the hypothesis in clear language. Also, the hypothesis should express something unambiguous enough that it is clear how you might go about testing it in order to confirm or disconfirm it. Consider this statement: "Childhood obesity is linked to the number of sugary drinks consumed daily." The observation being described here is, of course, childhood obesity, and we're given a suggestion as to one of its possible causes. Crucially, though, we're also given a statement that we can test in various ways. You might test it in an observational way, by comparing childhood obesity rates in populations that drink fewer sugary drinks to populations that drink more sugary drinks; or you might test it in an experimental way by designing a study in which you control as many other relevant variables as possible, to highlight the role that sugar consumption could play in childhood obesity.

By contrast, consider the central hypothesis of a book like Rhonda Byrne's *The Secret*, summarized as the "Law of Attraction." This "law" tells us (in life coach Bob Proctor's formulation): "Whatever is happening in your mind, you will attract it in your life" (De Fretes, n.d.). Is that hypothesis testable? No, because there is no way in which *it can fail to be true*. As hard as I (JR) might *think* I have massive wealth and fame "happening in my mind" right now, Byrne and her ilk can simply tell me that I'm not thinking hard enough, or not visualizing that wealth and fame in a sufficiently productive way. And then, of course, whenever good things might happen to me, they can still claim the credit because I *was* wishing it to be so. Books and movements like *The Secret* get to count their successes, and ignore their failures—the "law of attraction" is true in all instances because it isn't responsive to the *full* body of evidence (or any evidence at all) in the way that we would want a scientific hypothesis to be (Wheen, 2004).

Now, once a hypothesis is well enough established, we refer to it scientifically as a *law*. The "law of attraction" mentioned previously is an example of someone spuriously using the term "law" to give the illusion of credence to mumbo-jumbo, because, as we have seen, the hypothesis in question is not only not well established, but never could be. By contrast, the hypothesis that "things dropped from your hand will fall to the floor" is sufficiently well established that we feel entitled to treat it as a law, and in fact do so in calling it the "law of gravity."

Next, a scientific *theory* is something that tends to be broader in scope than hypotheses and laws—it is a set of well-tested and well-supported hypotheses and laws, which in combination explain many more events, and make many more predictions, than the hypotheses or laws on their own. Here, for example, we would all be familiar with the "theory of evolution," which contains many hypotheses on one related theme, namely, the idea that natural selection guides the development of living organisms (Coyne, 2010). This is very dissimilar to how many people use "theory" in everyday terms, which generally just means "an idea that I have" (such as "Well, I have a theory as to why Nancy and John broke off their engagement").

Finally, and returning to some of the discussion around what "knowledge" means from Chapter 1, when we speak of scientific *facts* we're not referring to things that we *know* to be true with 100% certainty but instead referring to conclusions that are confirmed to such an extent that it would be reasonable to offer provisional agreement, and unreasonable to deny agreement. As suggested earlier in this chapter, we can—and should—change our minds when new evidence comes along that is stronger than competing evidence we've had to date. This doesn't mean that the facts themselves change—it instead means that we were simply wrong about what the facts were, and have had the opportunity to learn about that error.

SCIENTIFIC REASONING

We use scientific reasoning every day of our lives, whether or not we're doing so deliberately. When you hear the weather report calling for rain, but look out of the window to confirm whether or not it is raining, you're using scientific reasoning in the very basic sense that you're *testing a claim* against the evidence. Before we talk about the process by which we do so, let's look at a (slightly) more complicated example, which demonstrates certain key steps in the process of scientific reasoning. Those steps can be stipulated as:

1. Identifying a problem, or observation in need of explanation
2. Gathering information about the problem or observation
3. Formulating explanations (hypotheses) regarding the problem or observation
4. Conducting tests or experiments to see which, if any, of the hypotheses provide a resolution for the problem or explain the observation

5. Deriving a conclusion that accurately captures the resolution or observation (and, ideally, gives us guidance in terms of this or relevantly similar situations in the future)

An Example

Let's say that you're trying to charge your mobile phone, but the screen tells you that the phone is not charging, even though you confirm that you've plugged it in (and note that in confirming this, you've *already* started reasoning scientifically, by testing for the most obvious explanation). So here we're at Step 1: We have a problem or situation in need of explanation.

Step 2, information gathering, already started when you checked that you had plugged the phone in correctly. Step 2 might also involve elements like thinking back to when you last charged your phone. Everything seemed to work fine on the previous evening, so you know that something has changed, or gone wrong since then—now you just need to figure out what that is.

So (Step 3), what could explain the fact that it's not charging? Well, the power could be out in your neighborhood (that's one hypothesis), or just in your house (a competing hypothesis). Perhaps the problem is even more localized, and the power outlet you're using is faulty. The phone's charging unit or cable could also be a problem, or (and you don't really want to think about this one!), your $500 phone might be faulty.

You already know what to do in Step 4 because, as I say, we engage in this sort of reasoning process all the time, even if often unconsciously. You might be able to test the "power out in the neighborhood" and "power out in your house" hypothesis simultaneously, through realizing that you can hear the television playing in the next room. Then, you might rule out the power outlet being faulty, because your laptop is still receiving power despite being plugged into the same outlet. So now (in our stripped-down example), you're left with two hypotheses, and you'd probably next try testing a different cable—if the phone charges, you know the cable was faulty (which gives you Step 5).

If the phone still doesn't charge, you might think that this tells you the phone is faulty (also an answer to Step 5), but not quite yet—the second cable might also be faulty. So a possible next step would be to test both of these cables with a different phone to verify that they are able to charge a different device. It might just (phew!) be that all the cables you have available at the moment are faulty, and that there's nothing wrong with your phone at all.

This example uses the building blocks of hypotheses and evidence to reach conclusions, and is a completely familiar scenario to us because we do it or something close to it every day. But these building blocks can fit together in various ways as we go about the business of scientific thinking. The two most common ways of generalizing the process of scientific reasoning and how it proceeds are the methods of *induction* and *deduction*, both of which merit further discussion.

INDUCTION AND DEDUCTION

Our past experiences often form the basis of our expectations in the future. If you've had a couple of good meals at a particular restaurant, you'd start recommending it to friends because you expect that both you, and they, are likely to have good meals there in future. In other words, specific observations in the past are used to derive a general principle, which involves events in the future. This, in short, is *induction*.

The restaurant example, drawn from everyday life, is generated by a process of induction and works in exactly the same way in scientific reasoning—it amounts to the derivation of a general principle, even a law, from specific observations (Holland, Holyoak, Nisbett, & Thagard, 1989). Notice that for these observations to generate reliable principles that can accurately predict events in the future, we need to make an important assumption, namely, that nature is (to a significant extent, at least) *uniform*. On that assumption, we are identifying regularities and using those as fairly stable or consistent variables for understanding the world around us and for predicting future events.

How are these general principles justified? First, the number of observation statements must be fairly large for us to be able to have confidence in them not being outliers or accidents. Second, the observations must be repeated under a wide variety of conditions. To go back to our everyday example, the more meals *you* eat at a particular restaurant, the more strongly justified *your* conclusion is that it's a good restaurant. And second, if a number of different people (with different tastes and standards than you) have eaten at that restaurant at various times (in season, out of season, for lunch and dinner, and so forth) and also enjoyed it, the strength of the conclusion is also positively affected.

But as many observation statements as we have—even when they have the sort of diversity described previously—the truth of an inductive conclusion can never be guaranteed. This is because we are predicting events in the future based on ones in the past, and all sorts of unknown or unpredictable variables could still intervene to make our conclusion false (Goodman, 1983). Inductive conclusions are therefore not *truth-preserving*, in that even though your observation statements might be entirely true, they could nevertheless generate an inductive conclusion that ends up making false predictions about the future.

By contrast, the method of *deduction* tends to be better at being truth-preserving, but this can come at the cost of being somewhat narrower in scope. This is because the conclusions of deductive processes are never more general than the observations or evidence used to generate those conclusions. Deduction works from the general to the specific (or top-down), whereas induction is reasoning that works from specific to the general (or bottom-up).

Induction is therefore often more suited to *developing* hypotheses, and deduction better for *testing* them. Take the example of visiting your doctor when suffering from a fever after having recently visited an area where malaria is common. By induction, your doctor could reason that it's more likely than it would normally be for you to have malaria, seeing as (a) you're exhibiting symptoms common to those who have malaria and (b) you've recently been exposed to risk of contracting malaria. But seeing as it's flu season also, and your doctor has seen a number of patients suffering from flu-related fevers during this week, she now has two competing hypotheses, both formed by induction. The way to tell them apart, and to find the real cause of your discomfort, is to use deduction—in this case, a blood test that will indicate the very *particular* detail of whether or not the relevant parasites have infected your system. From their presence or absence, we can *deduce* that you either have or don't have malaria, and then treat you accordingly.

VERIFICATION AND FALSIFICATION

The nature of induction means that any number of confirming instances can never actually *prove* a theory to be true (because you're generalizing about the future from a limited sample of past occurrences). Any one falsifying instance can *refute* a theory, in that we learn that the theory cannot accommodate all observations, and needs revision in order to do so (Popper, 2002). This can introduce a conflict between scientific thinking and our common understanding of what sorts of conclusions we should regard as reasonable in our day-to-day lives. When we reason about our own lives, we're inclined to make fairly confident predictions about the future based on what we've experienced in the past, and to do so by *verifying* our beliefs through confirmatory experiences—in fact, we might even be more inclined to restrict ourselves to testing the hypotheses we think will turn out to be true, rather than the ones we're not so sure of (Klayman & Ha, 1987). Furthermore, our personal anecdotes can be treated as reliable data, rather than subjective and possibly inaccurate information. And although it is true that the larger and more diverse the number of observational statements we have supporting a conclusion, the more likely it is to be true, a key element of science is the search for *disconfirming* instances rather than verification through more confirming instances. This is because any number of theories could perhaps explain the confirming instances, and it is thus more useful to try to rule out the weaker theories.

This importance of falsification (rather than verification) is a key aspect that differentiates scientific thinking from our more natural way of thinking. Robust scientific theories need to be open to *falsification*, and scientific theories should ideally offer us—in the specification of the theory itself—what might count as falsifying conditions for the theory. It's important to make a distinction between two separate (though closely related) concepts here: falsifiability *in principle* versus a hypothesis actually having been *falsified* in practice.

From the anthropomorphized perspective of the hypothesis itself, it is of course not a good outcome for a hypothesis to be defeated by the evidence. But, as a scientist, you would hope to approach the evidence objectively, and believe what is most likely to be true, regardless of personal loyalties and preferences. So, from the imagined point of view of knowledge and scientific progress, it's actually a *good* outcome that a hypothesis is falsified in practice, in that we learn about another way in which we were previously wrong about something, and are (sometimes) given a clue as to where the correct answer might lie. To put it another way, for a hypothesis to be falsified rules out one potential source of error and unmuddies the waters of inquiry (at least to some extent).

For a hypothesis to be falsifiable *in principle* asks us to consider a different issue, namely, that we need to be able to imagine ways in which a hypothesis is *potentially* falsifiable. We need to have some mechanism (even if it is a hypothetical one) for testing it, to see whether it can survive those tests. This is a key distinction between science and pseudoscience (as will be further discussed in Chapter 3), in that claims made in fields like astrology tend to be so nonspecific that you cannot even imagine a method of putting them to a test. There are no possible falsifying instances of a general claim like "December will be a good month for you professionally" because even if you lost your job in December, the astrologer could say that this proves her point—you were, after all, stagnating in your current job, and are now free to find a different job, and one that can allow you to discover your *true* potential.

This tendency to look for *verification* of our beliefs, as the astrologer does in the preceding example, is intuitively sensible and it also has quite a heavyweight pedigree in the philosophy of science. In the 1920s and 1930s, the famous Vienna Circle developed the position known as "logical positivism," whereby a statement should be considered meaningful in only two sorts of situations: either it is a *formal* statement, such as you'd find in mathematics (2 + 2 = 4), or if it's capable of empirical verification (Sarkar, 1996).

However, as the astrology example shows, it's easy to find evidence that something is true. But finding confirming evidence doesn't tell us whether we were testing the correct hypothesis. The evidence could instead be confirming some other hypothesis that we haven't even considered yet (e.g., the hypothesis that we are superstitious and prone to believing all sorts of nonsense).

The Vienna Circle had the noble goal of trying to offer a clear principle to distinguish science from pseudoscience. They wanted to make it clear that metaphysical and theological statements were not cognitively or scientifically meaningful. But as Karl Popper pointed out in 1934 (Popper, 2002), the criterion of verifiability was too strong, and also too easy to satisfy. First, because some statements in science are useful and cannot (yet) be verified, like the ancient Greek notion of atoms could not be at the time; and second, because although it's relatively easy to verify a hypothesis, it's far more telling when a hypothesis is falsified, allowing us to rule it out as being true.

This is why the modern scientific method places a significant emphasis on attempts to falsify, rather than attempts to verify. To put it crudely, we take our various hypotheses, develop tests that would demonstrate those hypotheses to be false, and then see which hypotheses survive those tests. The ones that do are considered to be best justified and, in ordinary language, true (at least until replaced by a superior hypothesis, that is). We *triangulate on the truth* via eliminating falsehoods. It's also important to note another consequence of falsification: our conclusions are always provisional, in the sense that they haven't been falsified *yet*. New data could in future come to light that falsifies what we currently regard as true.

CONCLUSIONS

Figure 2.1 summarizes how, ideally, scientific thinking happens. You move from information gathering and hypothesis generation through inductive and deductive processes to help build ever-improving theories about both how the world is and why it is that way. A long and complex process, science continually puts its ideas to the test. Using this kind of method, we give up *certainty* in exchange for increased *confidence* that our beliefs are the best justified ones currently available to us, in full knowledge and humility that we might be getting aspects—or all of—the story wrong at the present moment. Pseudoscience, by contrast, starts with an inflated, and unwarranted, confidence in its conclusions being correct—and then proceeds in a manner that is completely immune to being corrected, because nothing can ever prove it wrong. This is the focus of our next chapter.

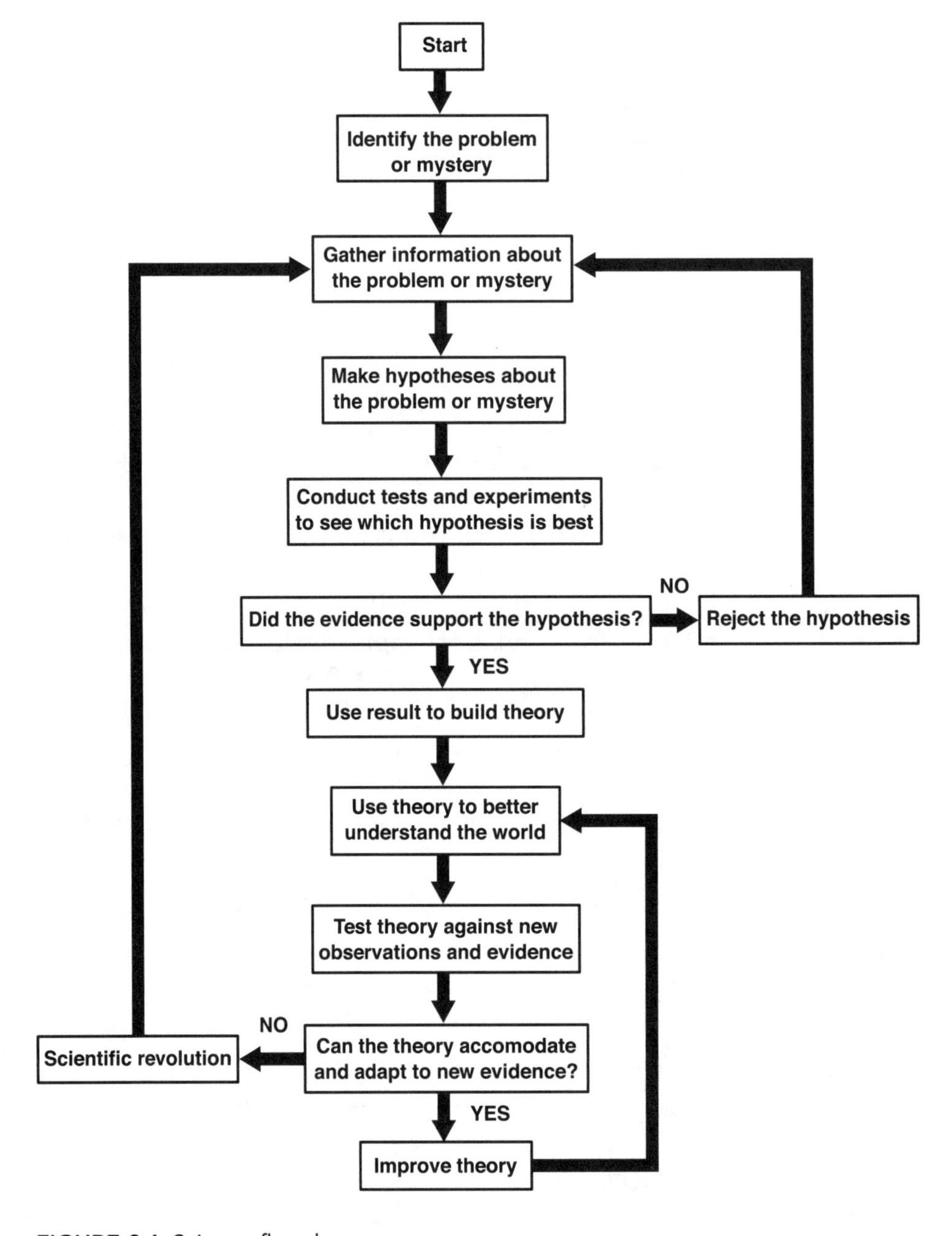

FIGURE 2.1 Science flowchart.

QUESTIONS FOR REFLECTION

1. *What do you regard as the most obvious difference between scientific thinking and our everyday ways of resolving questions? Should we always try to think scientifically about everyday problems? Why or why not?*
2. *Can you imagine ways in which the death of the traditional newsroom, with dedicated science editors and journalists, could positively impact the reporting of scientific developments?*

3. *If a belief cannot be investigated using the tools of science, can we be confident that it's a false belief? Why or why not? Can you think of some examples?*
4. *Do you ever consider whether your personal beliefs can be falsified? If not, do you now think that should you do so?*

REFERENCES

Adams, J. (2013). Collaborations: The fourth age of research. *Nature, 497*, 557–560.

Chabris, C. (2011). *The invisible gorilla: How our intuitions deceive us*. New York, NY: Harmony.

Coyne, J. (2010). *Why evolution is true*. London, UK: Penguin.

De Fretes, F. (n.d.). The secret *by Rhonda Byrne*. Retrieved from http://www.livinggood.com/book-reviews/bookreview-secret-rhonda-byrne

Deutsch, D. (2013). Why it's good to be wrong. *Nautilus, 2*. Retrieved from http://nautil.us/issue/2/uncertainty/why-its-good-to-be-wrong

Fahy, D. (2015). *The new celebrity scientists: Out of the lab and into the limelight*. Lanham, MD: Rowman & Littlefield.

Goldacre, B. (2012). *Bad pharma: How drug companies mislead doctors and harm patients*. London, UK: Fourth Estate.

Goldacre, B. (2014). *I think you'll find it's a bit more complicated than that*. London, UK: Fourth Estate.

Goodman, N. (1983). *Fact, fiction, and forecast*. Cambridge, MA: Harvard University Press.

Holland, J. H., Holyoak, K. J., Nisbett, R. E., & Thagard, P. R. (1989). *Induction: Processes of inference, learning, and discovery*. Cambridge, MA: MIT Press.

Ioannidis, J. (2005). Why most published research findings are false. *PLoS Medicine, 2*(8), e124. doi:10.1371/journal.pmed.0020124

Klayman, J., & Ha, Y. (1987). Confirmation, disconfirmation, and information in hypothesis testing. *Psychological Review, 94*(2), 211–228.

Lilienfeld, S. O., Lynn, S. J., Namy, L. L., & Woolf, N. J. (2013). *Psychology: From inquiry to understanding* (3rd ed.). New York, NY: Pearson.

Mlodinow, L. (2009). *The drunkard's walk: How randomness rules our lives*. London, UK: Vintage.

Okasha, S. (2002). *Philosophy of science: A very short introduction*. Oxford, UK: Oxford University Press.

Popper, K. (2002). *The logic of scientific discovery*. London, UK: Routledge.

Sarkar, S. (Ed.). (1996). *The legacy of the Vienna circle: Modern reappraisals*. New York, NY: Garland.

Shermer, M. (2002). *Why people believe weird things*. New York, NY: Holt.

Shermer, M. (2008). Patternicity: Finding meaningful patterns in meaningless noise. *Scientific American 299*(6), 48. Retrieved from http://www.scientificamerican.com/article/patternicity-finding-meaningful-patterns/?page=1

Wheen, F. (2004). *How mumbo-jumbo conquered the world: A short history of modern delusions*. London, UK: Harper.

CHAPTER 3

WHAT IS PSEUDOSCIENCE?

Metaphysical claims involving things like the "Law of Attraction" (Byrne, 2006), astrology, or homeopathy all share at least two features, both of which were identified as a problem in Chapter 2: it's very easy to find evidence for those claims, and any claims made can typically not be falsified. The broad set of fields for which this holds true are often collected under the terms *pseudoscience* or *junk science*.

Pseudoscientific claims make predictions or offer explanations, just as scientific claims do, and so can be difficult to distinguish from "proper" scientific claims. Where they differ is in the scaffolding for the claims made: Pseudoscientific theories fail to offer a robust set of underlying laws, or even hypotheses, that can be empirically shown to justify the predictions or explanations generated by those theories. Pseudoscience is often taken seriously, and thought to correspond to reality, for quite mundane reasons—our gullibility and will to believe. Consider astrology, which claims that the zodiac sign under which you are born influences your personality and can even help to foretell your future.

A psychologist named Bertram Forer showed how easy it is to tap into the human frailties of wanting to believe and gullibility in a landmark experiment some 70 years ago. Forer administered a personality test to his undergraduate students in one class period and promised to give them personalized feedback during the next class. But rather than giving them individual assessments, he copied a few descriptive sentences from a newspaper astrology column and gave all the students the exact same "profile" (Forer, 1949, p. 120), reproduced as follows:

> You have a great need for other people to like and admire you. You have a tendency to be critical of yourself. You have a great deal of unused capacity which you have not turned to your advantage. While you have some personality weaknesses, you are generally able to compensate for them. Your sexual adjustment has presented problems for you. Disciplined and self-controlled outside, you tend to be worrisome and insecure inside. At times you have serious doubts as to whether you have made the right decision or done

> the right thing. You prefer a certain amount of change and variety and become dissatisfied when hemmed in by restrictions and limitations. You pride yourself as an independent thinker and do not accept others' statements without satisfactory proof. You have found it unwise to be too frank in revealing yourself to others. At times you are extroverted, affable, sociable, while at other times you are introverted, wary, reserved. Some of your aspirations tend to be pretty unrealistic. Security is one of your major goals in life.

When the students were asked to evaluate the accuracy of the character traits identified in their horoscope on a scale of 0 (least accurate) to 5 (most accurate), an average result of 4.26 was obtained. This result has proved to be replicable across various cultures and hundreds of repetitions in other classrooms since Forer ran the experiment. The average score remains at around 4.2 across these different contexts (Dickson & Kelly, 1985). What the Forer effect (also sometimes referred to as the Barnum effect, after the circus showman P. T. Barnum) shows is that we tend to accept highly generalized descriptions of this sort as accurate. We take notice of, and overvalue, apparently confirming instances of apparently plausible hypotheses, and discount or ignore evidence that runs contrary to what we're invested in believing.

WHY IS THIS A PROBLEM?

Although some versions of pseudoscience, mysticism, and general quackery are fairly constant insults to our sensibilities (as we will see in Chapter 11, some forms of alternative medicine have been around for hundreds of years), others, such as those magic holographic bracelets that promise you increased balance and strength, seem to go in and out of fashion like pop stars and children's toys. Although it is not always clear that these fashions cause direct harm to our health, they often cause at least two sorts of indirect harm. The first sort of harm occurs through quackery taking the place of effective medical interventions, thereby allowing people to suffer needlessly. Sometimes, people even die, as was the case with 9-month-old Gloria Sam, an Australian infant whose (treatable) eczema became chronic and—through infection—deadly after her homeopath father chose to "treat" her with water instead of medicine (Associated Press, 2009).

The second sort of harm is to our wallets, in that mystical interventions always come at a price. Sometimes you might consider investments worthwhile even if they are for things that are not directly tangible or to your benefit—the 10% that some religious folk tithe to their churches does at least support various forms of communal activity, regardless of the truth or falsity of the religious beliefs. It's perhaps true that the gullible don't always deserve protection from their poor judgment—if it comforts you to build an airport for aliens, as the town council of Arès in France did in 1976 (*Daily Telegraph*, 2010), the ensuing merriment for the rest of us could well be worth the cost, assuming the money should not instead have been spent on basic sanitation or other needs.

But not everyone has money to waste on such follies. For every person with more dollars than sense, there will be plenty who might instead believe that some mass-marketed trinket can bring them increased health, wealth, or happiness, and who then proceed to allocate resources they cannot afford to purchasing those trinkets. Instead of choosing to take (expensive) medication, those suffering from bursitis or arthritis could choose to buy a Power Balance bracelet at a relatively cheap price (of roughly $30)—a one-off expense instead of a lifetime of medication.[1]

The claims made on behalf of these bracelets are absurd, and create a clear (yet misleading) impression that it's more than simply a placebo effect at work in referring to "Eastern philosophies [that] contain ideas related to energy. These are commonly referenced as Chi or Chakras. There are a number of well known practices like acupuncture, meditation and Feng Shui, which are believed to affect these energies" (Power Balance, n.d.-a).

Power Balance—similarly to all manufacturers of such products—does cover its legal bases while making vague concessions to the fact that they know they are simply selling a more modern form of snake oil—this time in colorful silicone form. In answer to the question "How do I know it is working?" for example, they say, "Wear the product throughout your day, whether exercising or not. While we have received testimonials and responses from around the world about how Power Balance has helped people of all ages and physical abilities, there is no assurance it can work for everyone" (Power Balance, n.d.-b). Given the short memories people seem to have of these sorts of exploitation, this strategy should work for long enough to make a fair amount of money. They surely know, for example, that the manufacturers of the Q-Ray bracelet in the United States lost a class-action suit in 2006, and that the judge forced them to pay back $22.5 million of their "ill-gotten gains" (Federal Trade Commission, 2008). And that Power Balance in Australia was required to publish "corrective advertising" and refund all customers who asked for their money back (Australian Competition and Consumer Commission, 2010).

But our memories are short indeed, and we also often fail to do our homework when something new and apparently wonderful comes along. We forget, or never learned, that the demonstration a Power Balance salesperson will perform for a prospective buyer is a simple variant of what is

[1] Well, perhaps once-off: some of these magical bracelets, even the "quantum" sort, apparently need occasional recharging to replenish their powers.

known as "applied kinesiology," (Schwartz et al., 2014) frequently used by stage magicians to produce entirely subjective perceptions of increased strength and balance. We allow ourselves to forget—in the hope of finding some magic bullet for health and happiness—that we have no evidence for the existence of an "energy field" in humans that can be affected by negative ions or biofeedback (see Chapter 12 for more on that topic). We allow ourselves to ignore the fact that we don't know whether holograms can be "imprinted" with frequencies, how this might be done, and, if it can be done, on what principles the manufacturers choose the frequencies or amount of mysticism to cram into their bracelets. Most of all, we allow ourselves to forget that if you think about it for a moment, all of these possibilities seem vanishingly unlikely to be an accurate description of what's going on.

If you take a look at the "research" described on the web pages of the manufacturers of these products, you'll quickly notice that none of them conform to the commonly understood gold standard for scientific research, namely, the double-blind controlled trial (discussed in detail in Chapter 11). Instead, they consist of strings of meaningless technobabble and fancy words like "quantum," along with user testimonials. But user testimonials are no more than anecdotes, and no matter how many anecdotes you might accumulate, they do not add up to scientific data.

The increased wealth of these snake-oil peddlers comes at a broader cost than simply offering false hope to a few, and the waste of money that buying these bracelets entails. They contribute to us not doing our homework, and perhaps becoming more gullible and more ignorant. They tap into the absurd deference we afford to celebrities and their product endorsements, regardless of the fact that many of these celebrities have no relevant expertise (and that they often seem no more successful for wearing magic bracelets themselves).

This general "climate of unreason" is a dangerous thing in itself, regardless of the triviality of any particular *instance* of unreason. Power Balance, T4ProBalance, Quantum Leap, and all the other variants of these pyritic placebos can assist in raising our tolerance for quackery in general, and quackery comes in forms far more dangerous than a hologram. It was quack science that led to years of HIV/AIDS denialism in countries like South Africa, and quack science that supported, and still supports, conspiracy theories about vaccines causing autism (Harris, 2010).

It would be foolish to make any claims for a causal link between our tolerating relatively benign forms of pseudoscience and the more dangerous ones described previously. But it is in environments where we refrain from being critical of nonsense that people develop misguided ideas, and it is therefore not implausible to suggest that we should take care to foster a more critical environment, even in cases like these, by applying scientific modes of thought and inquiry. This can help us to distinguish between pseudoscience and real science, regardless of the varying degrees of pseudo in some "science."[2]

THE DEMARCATION PROBLEM

Finding the boundary or differentiating criteria between science and pseudoscience is known as the demarcation problem, and, as is often the case in the philosophy of science, this problem has proven to be a thorny issue, with no clear consensus emerging despite years of debate. As Lakatos (1980, p. 1) reminds us, strength of conviction is no help at all:

> But the history of thought shows us that many people were totally committed to absurd beliefs. If the strengths of beliefs were a hallmark of knowledge, we should have to rank some tales about demons, angels, devils, and of heaven and hell as knowledge. Scientists, on the other hand, are very skeptical even of their best theories. Newton's is the most powerful theory science has yet produced, but Newton himself never believed that bodies attract each other at a distance. So no degree of commitment to beliefs makes them knowledge. Indeed, the hallmark of scientific behavior is a certain skepticism even towards one's most cherished theories. Blind commitment to a theory is not an intellectual virtue: it is an intellectual crime.

Despite the demarcation problem, a common-sense definition of pseudoscience is perhaps fairly easy to agree on—it's in the detail, especially in terms of what sciences fall into which category, that disagreements are likely to arise. A starting point for a definition would, we propose, look something like this:

> Any claim, hypothesis, or theory that is presented in the language and manner typical of scientific claims, but that fails to conform to accepted standards in science regarding openness to peer review, replicability, transparent methodology, and the potential for falsifiability is highly likely to be a pseudoscientific claim, hypothesis, or theory.

To put it another way, we are making the point that the *absence* of the hallmarks of good science, as described in Chapter 2, are a clear warning sign for the presence of pseudoscience. With pseudoscientific claims, anecdotes might be allowed to trump evidence, or confirmation (rather than falsification) might be given undue weight. Most important, perhaps,

[2] Much of the discussion here regarding Power Balance bracelets previously appeared in my (JR) column in the Daily Maverick of January 27, 2011.

the communal nature of the scientific endeavor is perverted—instead of making data, findings, and methodology open for all to scrutinize and attempt to replicate, pseudoscientific claims tend to rely on fringe data, published by fringe scientists, in low-quality journals, thereby creating an impression of scientific controversy where none actually exists. At the conclusion of Chapter 2, we presented a flowchart of science, showing, ideally, how it works. Figure 3.1 shows the analogous flowchart for pseudoscience.

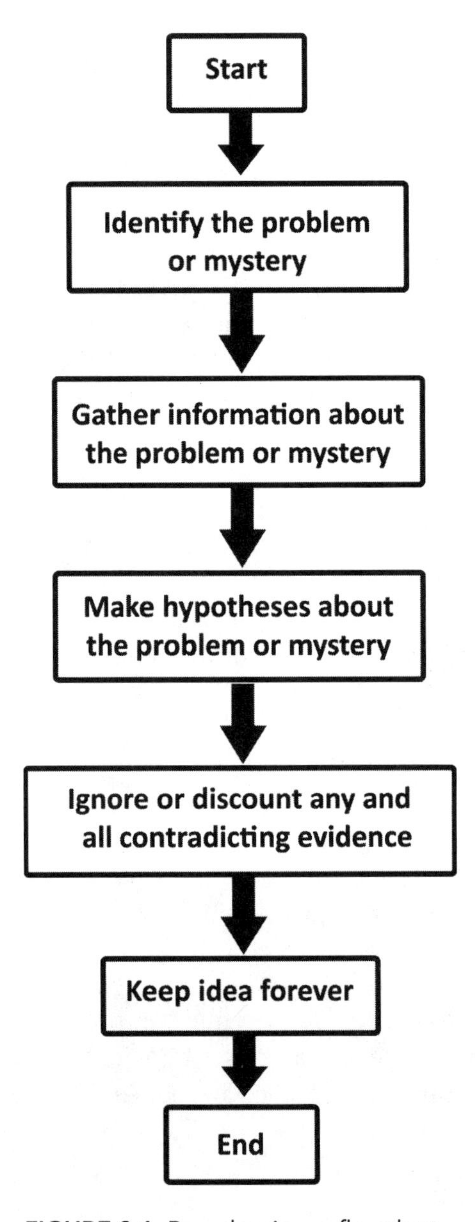

FIGURE 3.1 Pseudoscience flowchart.

Although slightly tongue-in-cheek, this flowchart shows how pseudoscience lacks the deductive, replicative, and corrective mechanisms that make science such a powerful explanatory force for finding out how the world works. In the real world, making the distinction between science and a pseudoscience can involve a difficult, and sometimes subjective, judgment. But even though it can be difficult, it's a judgment we should attempt to make whenever possible, as certain fields—perhaps particularly health care—are very prone to abuse by quack science, and real harms can result from public misunderstanding of the proper scientific method (Sagan, 1996).

Two final confounds need to be addressed before we take a closer look at some strategies for how to navigate (even if not completely resolve) the demarcation problem. First, it's not reliably the case that pseudoscientists or quacks are *intending* to deceive. In fact, a large number, maybe even the majority, might be utterly sincere rather than simply attempting to make some easy money off our gullibility. So as much as we might decry the *effects* of pseudoscience in terms of the harms it can do to people's wallets or, more worryingly, their health, we should remember to separate the issue of the scientific errors from our analysis of moral culpability and guilt. Although we might be able to be sure that the quack is wrong, we can't always be sure that he *knows* this, and intends to deceive. This would of course not make him right—but in one sense, it might make him (slightly) less wrong.

Second, there's a difference between pseudoscience or quackery and science simply done badly, or communicated incompetently. Take addiction as an example: Since 2007 or so numerous articles in the popular media have asserted that sugar is "addictive," that rats prefer it to cocaine, and so forth. The claim has been repeated so often that many people now uncritically accept it as true that sugar is in fact "addictive" (Kirkpatrick, 2013) rather than simply something we like the taste of, sometimes eat too much of, yet are able to stop eating too much of with a little more self-discipline (just like, say, watching TV or playing video games).

We'll look at this example in greater detail in Chapter 6, but for now, consider a recent newspaper article warning of the "hidden sugar in sauces" (Koen, 2015). For the sake of the example, just imagine we are agreed (a) that added sugar is something that should be consumed in moderation, if at all and (b) that accurate food labeling is a good thing, in that we want to know what we are eating. Even so, some things make for illegitimate comparison. Noting that a particular chili sauce contains four times the amount of sugar per 100 grams than a tin of Coca-Cola is profoundly misleading, thanks to the fact that a typical portion of the sauce might be something like 20 milliliters whereas a typical portion of the soda would be at least 10 times that amount, and usually more than 20 times that amount.

This is not an example of pseudoscience in the sense that something mystical is being measured, or weird and wonderful effects are alleged to result from some particular intervention. It is nevertheless misleading, and troubling as a result, but mostly as an index of a media and society that are attuned to taking scare stories more seriously than they should.

TIPS FOR IDENTIFYING PSEUDOSCIENCE

There are no hard-and-fast rules for distinguishing science from pseudoscience. In fact, one key problem in this regard is that pseudoscience trades exactly on these ambiguities and difficulties. Because many of us aren't comfortable with uncertainty, pseudoscience has a certain allure in presenting unambiguous "truths" or guidance to an audience. But as we argued in Chapter 2, the world of empirical data is messy, and the language of certainty will typically involve at least some misrepresentation of a more complex picture. But although assessing whether science is credible or not does involve a certain amount of subjective judgment—and making these judgments is a skill that can be exercised and refined—there are some rules of thumb that we hope you will find useful. The features highlighted next tend to be associated with pseudoscience more often than they are associated with good science, and that allows us to treat them as warning signs for claims that we are justifiably suspicious of.

Sensationalism and Oversimplification

Your first and most general clue involves encountering research that is presented in a sensationalistic or overly simplistic sort of way. You'll often encounter this right in the headline, where it might appear that a complex scientific story is being summarized in the form of "clickbait"—in other words, phrased in such a way that you're more likely to read it, whether or not it's accurate. "Shocking findings reveal that your shampoo is giving you cancer!" would be the sort of thing to watch out for, in that even if you know nothing about the study in question, you do know enough about the scientific method that (at the very least) you'd expect to read that your shampoo *might be* giving you cancer. Even this more modest claim is probably false, but at least the headline builds in some acknowledgment of the probabilistic nature of inductive reasoning, as well as the possibility of later falsification.

To return briefly to the earlier "sugar addiction" conversation: It merits highlighting here that even though our previous treatment of the topic presented it as poor communication of scientific activity rather than as pseudoscience, it does present a good example of how those boundaries can easily blur. If you, as a health care practitioner, present as axiomatic the claim that sugar is *known* to be toxic and/or addictive, and base your treatments on that "knowledge," it would certainly be reasonable to wonder whether misrepresenting the scientific consensus to such an extent that charges of making pseudoscientific (rather than merely exaggerated) claims might be a reasonable accusation.

Press Releases, Jargon, and "Churnalism"

Beware of "scientific" language or excessive jargon. We've recently seen a large upturn in the use of what has been described as "neurobabble" (or "neurobollocks"), which refers to the trend of presenting any human activity as being describable in terms of the brain and fMRI data (Poole, 2012). Although brain science holds enormous potential for answering some very important questions around things like addiction, it also allows us to escape personal responsibility—we might be encouraged to start thinking, "It wasn't me, it was my brain!" (Satel & Lilienfeld, 2013).

In an effort to sound impressive, press releases will often bombard readers with technical jargon because it instantly creates the impression that some smart people have done some very important work that you're unlikely to ever understand, so you basically just have to trust them. But this isn't true for two reasons. First, because there are most often nontechnical ways to describe the same thing—in fact, this is a hallmark of good popular science communicators. Second, because even if you don't understand various technical details, there are still common elements to the scientific process (e.g., best practice in study design) that you *can* understand and use to establish whether you should be inclined to trust the research in question or not.

Another element of oversimplification is found in the fact that science journalists can be guilty of taking a press release and repurposing it (or simply republishing it) without ever reading the actual study in question, which can result in a very incomplete and misleading picture being communicated to the public. This practice is (unfortunately) common enough that it even has a name: "churnalism." As a result of this journalistic laziness, websites have emerged—but sadly, mostly failed to survive—dedicated to helping you discover instances where instead of writing original copy, journalists simply tell you what a public relations company hoped they would tell you.

As Robbins (2011) has persuasively argued, there are two clear objections to churnalism. First, it undermines editorial integrity in that readers assume they are reading objective journalism, but are instead simply reading an entirely uncritical—in fact, partisan—summary of a study or series of events. Second, churnalism threatens the livelihood and value of paid journalists, whether in science or elsewhere. We value good journalism precisely because we trust that someone has done the research and can offer us insight into whether to trust research findings or not. The work isn't easy, and can require a level of expertise that laypeople typically lack. This expertise could be lost unless we can encourage writers to enter the field and excel at it—and this would in turn be impossible unless we remunerate such writers fairly.

Of course, it might be too late to think this battle can be won—science editors are a dying breed in newspaper offices, never mind science journalists, so the lesson is perhaps simply, and unfortunately, that science reporting in the popular news media is quite often anything but scientific. Even more worrying, this practice is not limited to journalists being irresponsible—a recent paper in the *BMJ* argued that a full 40% of press releases already contain exaggerated advice, meaning that what the journalists are regurgitating is a source that was already fairly unreliable (Sumner et al., 2014). So, whenever you can, read the original study rather than the press-release version of it.

Conflicts of Interest

There are some situations in which we are well aware that someone might be trying to make a fool of us or exploit our trust for financial or other gain. Think of the prototypical used-car salesman, for instance. The exploitation doesn't need to be particularly strong or overt for this to be the case, and it's not even consistently malicious. A salesperson might not be trying to sell us a *bad* or *defective* product, but nevertheless, she might be trying to sell us one particular product rather than another for the sake of a higher commission. The same reasoning can easily apply in scientific research: When a scientist is quoted or otherwise seen to be endorsing a particular drug or line of inquiry, it might be the case that he is employed by the company marketing that drug, or might have some other incentive to see it succeed, such as stock options or nepotistic connections. These sorts of relations don't *necessarily* corrupt or invalidate research, but they are worthy of concern in that they might do so, and are certainly factors that need to be considered when weighing any endorsement.

As ever, the judgment call regarding when to consider research compromised is not an easy one—and our own biases can get in the way of making a sound judgment. For example, when a scientist is employed by an advocacy group for a cause we support—hypothetically, let's say the World Wildlife Fund, with research involving the protection of some rare species—it should still be the case that we consider whether there is a conflict of interest, and therefore, whether the research in question is potentially compromised (Kiem, 2007). In a shift from even 15 years ago, most scientific journals today require the authors of articles they publish to list potential conflicts of interest and funding sources in order to increase the transparency of this issue.

Anecdotal Evidence

Anecdotal evidence consists largely of entirely uncontrolled observations, such as a person or group of people reporting that something happened to them, and others have no way of establishing the truth of their claim. This is not to say that they are lying, because lying is only one way of being wrong—they might simply be mistaken, perhaps fooled by their brains (see Chapter 5) or their world (see Chapter 6). For example, I (JR) might report that I experience greater levels of energy and alertness when consuming a particular herbal remedy, which I assiduously add to my morning tea. But that "data" is of little use unless we are sure that I'm not *also* sleeping better, or exercising more, or eating more healthily. This is why scientific studies explicitly control for these other variables, in order to give us greater confidence that it is the herbal remedy, rather than other factors, that are giving us the result we're observing.

Small and/or Unrepresentative Sample Sizes

The issue of sample size is related to the problems with anecdotal evidence, in that even if a study is controlled for known variables of the sort described previously, without testing that herbal remedy on a sufficiently large number of subjects, we can't know whether the (small) group we did test it on happen to share a common characteristic we didn't think of, and therefore didn't control for. A large sample size increases randomness, and therefore introduces such a diverse range of secondary factors (diet, sleeping habits, unanticipated factors, and so forth) that we can achieve a higher level of confidence that our observation regarding the remedy's efficacy will be less likely to be the result of chance, and more likely to point to some genuine causal relationship.

A familiar and useful analogy is that of elections. We value a high voter turnout because if 50% of potential voters actually vote and express a 70% confidence in the leadership of Sarah Jones, we have a far stronger indication of the community in general supporting her than if only 10% of voters express a view. The larger voting population is also unlikely to be homogenous, and this is the value of trying to ensure a *representative* sample. Even if a sample is large, the quality of evidence is impaired if the participants are all similar in some crucial way (and we might not know, in advance, what is and isn't crucial). Voters in a general election should not all come from the same economic class, for example, because people with similar class interests may well be more sympathetic to one political platform over another—better schools for their children might be a greater concern than crime if the voters tend to live in relatively crime-free areas.

Cherry-Picking

As we pointed out in Chapter 2, many areas of scientific inquiry involve debate and controversy. Despite this, it's frequently the case that strong results emerge in favor of one conclusion or another, and if you only look at studies that support one side of an argument, it's easy to get the impression that a consensus exists, and that "all the evidence" supports a particular point of view. This is the problem of "cherry-picking," and pseudoscience is rife with it. As Novella (2014) points out, it's even possible to find a glimmer of support for homeopathy (one of the most implausible of the alternative medicines; see Chapter 11) if you look hard enough, or if you interpret a study charitably enough.

But a glimmer of support, no matter how tenuous, doesn't resolve the debate, because it's the *totality* of evidence that should inform our attitudes and conclusions. A cherry-picker interprets the totality of evidence to be something closer to "the studies that happen to support my point of view"—and with that attitude, it should come as no surprise to us to find that they believe the evidence to be on their side.

No Control Group, No Blind Testing

Control groups are a standard feature in clinical trials examining whether a particular treatment helps a particular problem, because they allow us to eliminate many chance findings, or the possible effects of variables we didn't foresee in planning our study. Let's imagine that you want to test a new headache tablet—you already know it's safe (for the sake of argument), but you're not quite sure whether it *works*. To test the efficacy of this pill, it would certainly be helpful to give it to a (large and representative) group of people—but to make your findings even more reliable, it would be better to give a randomly selected group of people a different pill that looks and tastes just like the drug you're wanting to test but is chemically inert (in other words, a *placebo*). If you find that the control group benefits as much (or more!) than the group getting the "real" drug, you know that your drug doesn't do what you think it does.

In performing this test, it's important that people don't know whether they are receiving the real drug or the placebo, because that knowledge might in itself have an effect on how they respond. If you know you're getting the real drug, you might persuade yourself that your symptoms aren't as bad as they were a half hour ago and report that the drug is working. And if you're getting the placebo, you might perhaps think things are worse than they are, and report an excruciating pain. You'll notice that in both cases, you'd be reporting an anecdote, and a particularly unreliable one, because you've been *primed* to think in a certain way, based on knowing which group you're in.

It's even better if the researchers *themselves* don't know whether you are in the control group or not, because the ways in which they relate to you on the basis of that knowledge, might also have an effect on how you respond to the treatment or the placebo. When you don't know which group you're in, the experiment involves *blind* testing, and if neither you nor those who are directly involved in running the experiment know which group you're in, the experiment is being conducted under *double-blind* conditions (in Chapter 10 we will go over clinical trials and testing the effectiveness of treatments in greater detail).

THE LIMITS OF SCIENCE

It's perhaps easy to see the attraction of some pseudoscientific claims. Our own anecdotes are psychologically meaningful (to us), and the world of real science has difficult stories to tell, often resulting in no conclusive answer (yet). By contrast, pseudoscience allows room for those personal narratives to have meaning, and even explanatory power. If you don't like the sounds of what "scientists" are saying, or see some sort of conspiracy in what they say, you'll usually be able to find an alternative point of view that's more satisfying, even if it violates various principles of sound scientific reasoning.

But there's a conflict between our desire to have conclusive answers and what the scientific method is capable of. Science is less about final—or ultimate—answers, than about a way of evaluating claims and gathering the evidence that informs those claims. Science also cannot draw conclusions about things that fall outside of the empirical realm— things that we cannot measure or manipulate experimentally. This inability is not to be taken as proof one way or the other on any of these claims, but simply makes the point that at any given time there will be various claims that fall outside of the scientific realm. In time, as we discover more about the world around us, some or all of those claims might fall inside the net of things that can be scientifically investigated.

The point is that science is not omniscient, nor is it infallible. Science is conducted by human beings, who are known to be prone to making mistakes or misinterpreting information. But even so, it's simply the best method we have for understanding the universe, particularly given the error-correction mechanisms that are integral to it. To quote Albert Einstein: "One thing I have learned in a long life: that all our science, measured against reality, is primitive and childlike—and yet it is the most precious thing we have."

CONCLUSIONS

To some extent, anyone can be a scientist. This doesn't mean you that have to put on a white coat and perform formal experiments in a laboratory, but rather that anyone can understand the basics of what the scientific method looks like, as well as what the warning signs of pseudoscience look like. As long as we remember that a shortage of knowledge or information is never an excuse to simply make things up, or to dogmatically believe what you would prefer to be true regardless of the evidence that does exist, and instead test your ideas in a controlled way that helps to reduce bias, you are thinking like a scientist. You are employing, critical thinking, which is the focus of our next chapter.

QUESTIONS FOR REFLECTION

1. *Health care plans in some countries, including the United States, cover the costs of consulting with pseudoscientific practitioners such as professional homeopaths or chiropractors. Is this appropriate, given that payments in collective schemes like these affect everyone's contributions?*
2. *Imagine being presented with a belt that is said to create healing vibrations that ease lower back pain. How would you go about testing it to see whether it's a scam?*
3. *Should anecdotes ever be taken seriously in scientific inquiry? When do you think they could add value?*
4. *Quack science is fairly common—many newspapers carry astrology columns, for example, and you can often see people on television who claim to be psychics. Do you think publishers and broadcasters have a moral obligation to avoid publicizing quackery?*
5. *Do you agree with the idea that humans in general seem to dislike uncertainty and ambiguity? Why do you think this is the case?*

REFERENCES

Australian Competition and Consumer Commission. (2010). *Power Balance admits no reasonable basis for wristband claims, consumers offered refunds*. Retrieved from https://www.accc.gov.au/media-release/power-balance-admits-no-reasonable-basis-for-wristband-claims-consumers-offered

Associated Press. (2009, September 28). Homeopathy couple jailed over daughter's death. *The Guardian*. Retrieved from http://www.theguardian.com/world/2009/sep/28/homeopathy-baby-death-couple-jailed

Byrne, R. (2006). *The Secret*. Hillsboro, OR: Beyond Words Publishing.

Daily Telegraph. (2010, December). *First landing on 34-year-old UFO airport at Ares in France*. Retrieved from http://www.dailytelegraph.com.au/news/weird/

first-landing-on-34-year-old-ufo-airport-at-ares-in-france/story-e6frev20-1225914443251?nk=c5ef09fcafd589acf69be37dd20b7a6e

Dickson, D. H., & Kelly, I. W. (1985). The 'Barnum effect' in personality assessment: A review of the literature. *Psychological Reports*, *57*, 367–382.

Federal Trade Commission. (2008, July). *Appeals court affirms ruling in FTCs favor in Q-ray bracelet case*. Retrieved from https://www.ftc.gov/news-events/press-releases/2008/01/appeals-court-affirms-ruling-ftcs-favor-q-ray-bracelet-case

Forer, B. R. (1949). The fallacy of personal validation: A classroom demonstration of gullibility. *Journal of Abnormal and Social Psychology*, *44*(1), 118–123.

Harris, G. (2010, February 2). Journal retracts 1998 paper linking autism to vaccines. *The New York Times*. Retrieved from http://www.nytimes.com/2010/02/03/health/research/03lancet.html?_r=0

Kiem, B. (2007, December). When is a conflict of interest a conflict? *Wired*. Retrieved from http://www.wired.com/2007/10/when-is-a-confl

Kirkpatrick, K. (2013, November 19). Addicted to sugar? 7 steps you need to take before you can break free. *Huffington Post*. Retrieved from http://www.huffingtonpost.com/kristin-kirkpatrick-ms-rd-ld/sugar-addiction-_b_3861957.html

Koen, G. (2015, March 29). Seeking out the hidden sugar in sauces. *City Press*. Retrieved from http://www.citypress.co.za/lifestyle/seeking-out-the-hidden-sugar-in-sauces

Lakatos, I. (1980). Introduction: Science and pseudoscience. In J. Worrall & G. Currie (Eds.), *The methodology of scientific research programmes: Philosophical papers* (Vol. 1, pp. 1–8). Cambridge, UK: Cambridge University Press.

Novella, S. (2014). *The clinical evidence for homeopathy*. Retrieved from http://theness.com/neurologicablog/index.php/the-clinical-evidence-for-homeopathy

Poole, S. (2012, September 18). Your brain on pseudoscience: The rise of popular neurobollocks. *New Statesman*. Retrieved from http://www.newstatesman.com/culture/books/2012/09/your-brain-pseudoscience-rise-popular-neurobollocks

Power Balance. (n.d.-a). Power Balance–the originator of performance technology. Retrieved from http://www.powerbalance.com/aboutus.asp

Power Balance. (n.d.-b). Power Balance™–FAQs. Retrieved from http://www.powerbalance.com/faqs.asp

Robbins, M. (2011, April 27). Science churnalism. *The Guardian*. Retrieved from http://www.theguardian.com/science/the-lay-scientist/2011/apr/25/1

Rousseau, J. (2011, January 27). Opinionista: Dr. Woo and the silicon snake-oil bangle sellers. *The Daily Maverick*. Retrieved from http://www.dailymaverick.co.za/opinionista/2011-01-26-dr-woo-and-the-silicon-snake-oil-bangle-sellers

Sagan, C. (1996). *The demon-haunted world*. New York, NY: Ballantine Books.

Satel, S., & Lilienfeld, S. (2013). *Brainwashed: The seductive appeal of mindless neuroscience*. New York, NY: Basic Books.

Schwartz, S. A., Utts, J., Spottiswoode, S. J., Shade, C. W., Tully, L., Morris, W. F., & Nachman, G. (2014). A double-blind, randomized study to assess the validity of applied kinesiology (AK) as a diagnostic tool and as a nonlocal proximity effect. *Explore: Journal of Science and Healing*, *10*(2), 99–108.

Shermer, M. (2002). *Why people believe weird things*. New York, NY: Holt.

Sumner, P., Vivian-Griffiths, S., Boivin, J., Williams, A., Venetis, C. A., Davies, A., . . . Chambers, C. D. (2014). The association between exaggeration in health related science news and academic press releases: Retrospective observational study. *British Medical Journal*, *349*, g7015.

CHAPTER 4

WHAT IS CRITICAL THINKING?

All of us have likely been instructed to "think critically" about something before. Maybe it was in a classroom, by a friend offering advice before you made a big decision, or in numerous other settings. Unfortunately, few of us have actually ever been shown *how* to think critically about information. For reasons outlined thoroughly in the next two chapters, critical thinking does not come naturally to most of us. Instead, it is a skill to be learned, analogous to riding a bicycle or playing the guitar. This chapter offers guidance and advice to get you started down the road to becoming a critical thinker about anything you encounter.

WHAT IS A SKEPTIC?

When discussing various ideas related to critical or scientific reasoning and how to assess evidence, we shouldn't forget that we could be the victims of various biases ourselves. Furthermore, even those of us who identify as a scientific skeptic (shortened to just "skeptic" in the rest of the book) shouldn't be allowed to forget (as discussed in Chapter 5) that we can be prone to all the biases that are common to humans—we don't like to be wrong, and this leads us to develop ways of thinking and interpreting evidence that makes it seem (to us, at least) that we're wrong less often than the evidence suggests.

A simple definition of a skeptic might involve highlighting exactly this problem, as well as the need to do what we can to overcome it. A skeptic is led by the evidence, and tries the best she can to set aside the biases and preconceptions that lead us to error. Speaking of error, a skeptic is also willing to be wrong—she seeks out falsifying evidence, tries to not be overly credulous, and focuses strongly on the process and method by which true beliefs are more likely to result than false beliefs. This entails not only being willing to be wrong, but also being willing to say "I don't know" when the evidence is incomplete or unknown. This is an important attitude or style to cultivate, because as long as we are resisting unwarranted confidence or the appearance of it, we're signaling to others and reminding ourselves that the evidence still matters and that our minds can still be changed.

We're emphasizing this idea because our considered views are—or should be—always contingent on the information to which we have access. Therefore, we are often in a position to confess in advance that our information is inadequate for conviction to be a reasonable attitude. We might nevertheless sometimes feign conviction in conversation, partly because many of these debates take place on social media platforms that are intolerant of nuance. But we think that it's precisely the role of skeptics to fight for nuance, and to demonstrate, partly through showing that we're willing to embrace uncertainty, why we should be taken seriously when we claim that some conclusion or other is strongly justified. We devalue our skeptical credibility when we assert certainty, and do harm to the political cause of skepticism (which is, in short, using evidence to guide beliefs and decisions) in giving the impression that it is an alternative dogma, rather than an epistemological approach that respects and requires evidence.

Speaking of the politics of skepticism brings the history and current state of skepticism into the frame (Loxton, 2013). Some who identify as skeptics regard the label as signifying a default position of distrust with regard to any claim that one might encounter, to the extent that any consensus position might be regarded as suspicious. To put it simply, "skeptic" is sometimes misinterpreted as being synonymous with doubt or denial of a consensus narrative. For us, the focus is misplaced with these uses of the word *skeptic*. A skeptic would not dogmatically and automatically be suspicious of claims that meet the basic standards of scientific reasoning, that conform to common knowledge, that aren't made in service of suspicious motives, and so forth.

As Daniel Dennett, Bill Nye, and others point out in a statement released by the Committee for Skeptical Inquiry (CSI; 2014), skeptics and "deniers" are not the same thing, in that "AIDS deniers" typically refer to people who are "hyperskeptical" rather than skeptics—they distrust the standard narrative that HIV causes AIDS and engage in pseudoscientific reasoning to support their denialist positions. In the example offered by the CSI, the same pattern is demonstrated toward anthropogenic climate change, in that Senator James Inhofe's "belief that global warming is 'the greatest hoax ever perpetrated on the American people'" is an example of denial of the overwhelming evidence that climate change is real, rather than an example of skepticism as defined here. The CSI statement makes the important point that these two concepts—skepticism and denial—need to be kept separate, in that skeptics do valuable work that should not be polluted by association with the denialism of those folk like Inhofe, who replace the critical thinking of skepticism with dogmatism and cherry-picking. Inhofe and other deniers are so committed to a particular conclusion that they fail to do the work of responsible scientific reasoning.[1]

[1] As we will see in Chapter 13, there are likely a number of influences for Inhofe's denialism.

Another way of illustrating this distinction is to highlight the difference between what one might call scientific skepticism and pathological skepticism. Scientific skeptics are open to new ideas and challenging the consensus and will (ideally) happily change their minds when presented with good evidence that a change is merited. Although they are skeptical of fallacious appeals to authority (claims that something is true just because someone important said so), they nevertheless recognize that there is value in expertise, or that it's often the case that the consensus of experts is a more reliable source of knowledge than the views of the layperson.

In contrast, pathological skepticism is characterized by closed-mindedness and cynicism (rather than skepticism) with regard to claims that are made in support of a consensus point of view. Although we're all prone to confirmation bias—readily accepting evidence as good if it agrees with what we're hoping is true—pathological skeptics are also prone to a disconfirmation bias (Edwards & Smith, 1996). That is, they seek out (and sometimes give spurious credit to) evidence that disconfirms the standard or consensus narrative. Their primary purpose is to defend their "skeptical" conclusions, rather than to discard conclusions as a consequence, or result, of applying the methods of scientific skepticism.

The mistake pathological skeptics make is to invest too much in these conclusions and too little in the method by which they are reached. Our worth as (scientific) skeptics is not vested in conclusions, but in the manner in which we reach conclusions. Skepticism is not about merely being right. Being right—if we are right—is the end product of a process and a method, not an excuse for dogmatism (or pathological or hyperskepticism).

Sometimes we need to remind ourselves of what that method looks like, and the steps in that process, to maximize our chances of reaching the correct conclusion. Focusing simply on the conclusions rather than the method can make us forget how often—and how easily—we can get things wrong. As skeptics, we need to set an example in the domain of critical reasoning, and show others that regardless of authority or knowledge in any given discipline, there are common elements to all arguments, and that everybody can become an expert—or at least substantially more proficient—in how to deploy and critique evidence and arguments.

OPTIMAL, NOT PERFECT, DECISIONS

The beliefs that we have, or the concepts and categories by which we understand the world, are the lenses through which we view and interpret the world. They are therefore crucial in determining key aspects of our lives, most notably, perhaps, whether we're doing the best job we can both in gathering information and then using that information in the manner most effective for achieving our desires or goals. However, a basic problem with decision making is that human beings, as individuals, are faced with a very difficult, if not impossible, task. We are supposed to make rational decisions about our futures in the face of an infinite number of complicating factors, the most significant of which is that there are so many pieces of information that we should (ideally) incorporate into the decision-making process, and a large number of other facts that we could, given enough time, incorporate as well.

If a truly rational decision had to include all these factors, it is clear that no one would ever be able to make any rational decisions. We would all be standing around, frozen by indecision and unable to do much of anything. So, the standard of *perfect* rationality is an unreasonable standard to aspire to. One term used to describe this problem is *bounded rationality* (Simon, 1978), which expresses the idea that our rationality is bounded, or constrained, by factors such as the time available for thinking about a problem, the information we have access to, and the limitations to rationality that sometimes arise though a shortage of resources—as when financial constraints result in scientific research projects being terminated before they have been concluded.

In a general sense, any problem that needs to be resolved can be described as a research activity. This is because any problem or question is going to involve a series of steps that typically includes gathering information, arranging that information into a useful package, and then using that information to determine what the most appropriate course of action will be. All the steps in this process require making judgments: about which information is relevant; how to rank the relevance of various pieces of information; how best to present that information; what the likely effects of doing so will be; to what extent those effects should influence how and what you communicate to others; and so forth.

Being able to follow those steps in a better rather than worse fashion is important not only in rarified contexts such as the scientific laboratory. Instead, we use this sort of process in larger social contexts like trying to understand why the crime rate or inflation is high, but also in everyday situations like determining whether or not to buy a new pair of shoes. The credibility of the outcome or decision reached depends heavily on the way in which the investigation was conducted—not only the quality of the evidence, but also the process used in arranging that evidence to reach a conclusion. Perhaps more important for the average person, those of us who pay attention to clear reasoning tend to report "fewer negative life events" than those who do not (Butler, 2012).

We can sometimes make serious errors in terms of what we consider to be good or bad evidence, which points to the need to develop an understanding of best practice for how we gather information, and then how we use that information. This understanding, and the skills involved in exercising it, could broadly be described as the process of "critical thinking."

THE SPACE OF REASONS

The activity of thinking critically occurs in what one might call the "space of reasons" (Sellars, 1956, p. 299), meaning that any time there are serious issues at stake, we should expect to be asked for the reasons why we believe one thing rather than something else (e.g., you should not be surprised to explain why public health concerns dictate that parents should give their children the measles, mumps, and rubella [MMR] vaccine rather than not doing so).

Similarly, we should feel entitled to demand such reasons from other people. If an action or decision taken by someone else is likely to affect you, it is perfectly legitimate to require evidence that the decision or action has been thought through, and for you to be reassured that the decision was taken carefully, not simply made as a consequence of some bias or prejudice. Critical thinking describes the process by which we make decisions that are capable of being defended with evidence through a process that—assuming we all play fair—would be considered acceptable by any rational agent.

Understanding human interaction as occurring in the space of reasons implies the possibility of understanding conversations, debates, and attempts at persuasion as consisting of countless ongoing debates and research projects wherein each person presents evidence and reasoning that she believes should lead to a particular conclusion. Other people then introduce their own evidence to show how yours may be incomplete or misleading, and therefore how your conclusion might also be faulty. Alternatively, they would sometimes lead you to see that even though your evidence can't be faulted, you have packaged it in an illegitimate way, reaching conclusions that aren't well *justified* by the evidence that you have presented.

Argument and Argumentation

What is described here is the process of argumentation, which consists of argument and counter argument. Even though people typically understand "argument" to mean an aggressive or emotional battle, here we use it to refer to a particular methodology that provides us with the best possible chance—though rarely a guarantee—of arriving at a conclusion that best fits the evidence and, when we are choosing between alternatives that affect our welfare, offers the most advantageous consequences (Briggs, 2014).

Although we don't often spell out the steps of an argument in ordinary conversation, this doesn't necessarily indicate the *absence* of argument. Instead, it might be the case that it's not efficient to detail the argument because of how long it would take to do so, or because we have long since adopted some conversational or social norms whereby we fill in the blanks of the (hypothetical) argument.

As an example, consider what your thought process might look like—if laid out in pedantic fashion—while deciding whether or not to order a hamburger at a restaurant:

1. I am hungry and should eat.
2. I have eaten lots of beef this week, so maybe I should have fish.
3. But I really like hamburger.
4. Or maybe I should have pasta, because I've been feeling guilty about all these animals being killed for my pleasure.
5. But the hamburger is only $10; the other options are far more expensive.
6. Conclusion: I'll have the cheeseburger, please!

We've left out many potential steps here, but we hope the example suffices to show that because it might only take you a few seconds to go through the reasons for and against some decision, we wouldn't even notice that our trivial (leaving aside concerns related to point #4) decisions about what to eat require a process of argumentation, with *argumentation* understood to be the weighing of reasons for and against something, leading to what we think of as the most rational conclusion.

If even these mundane choices involve a background process of argumentation, we shouldn't be surprised to find ourselves making arguments, or being able to make arguments, any time we attempt to persuade or influence other people. And if the methodology of rational argumentation provides us with the best possible chance of reaching optimal decisions, we should then be concerned with learning and understanding how to apply that methodology, because some of the decisions we make are going to have far more serious consequences than whether or not to order a hamburger.

Although we cannot necessarily ever *guarantee* that any decision taken is the best possible one, we can at least reach the state of feeling justified that a particular decision is the *most* rational in light of the circumstances, and therefore more likely to be the optimal decision.

Arguments and Nonarguments

Not every interaction with other people involves arguments, and it's not always necessary for whatever we say to be a matter of logic or rationality. We can also express feelings, ask questions, and describe events, to give just three examples of different forms of communication. Another type of nonargument is an explanation, through which we are less trying to justify something as being true and more trying to explain how it came about, or why it is true.

Another way of putting this is to say that explanations are frequently appropriate in situations where people agree on the truth of a claim, but one or more of them fail to understand exactly why it should be regarded as true. It is important to note, though, that explanations often still look like arguments in terms of their structure. Consider the weather report on the news:

1. There was a high-pressure cell over Los Angeles yesterday.
2. High-pressure cells of that nature result in rain roughly 90% of the time.
3. So it rained in Los Angeles yesterday.

In this example, it was never in dispute that it rained in Los Angeles yesterday, so that "conclusion" didn't require an argument to justify it. Notice, though, that this explanation could quickly become an argument in at least two cases. First, the weather report could be *predicting* instead of explaining rain, in which case an argument (to be specific, an inductive–statistical argument) is being presented; or alternatively, someone could dispute the evidence presented in 1 or 2, in which case an argument would need to be presented showing that these two pieces of evidence are acceptable.

Arguments should not be thought of as being necessarily better or worse than any other sort of communication, but we should understand that they serve a particular purpose, and that operating in the space of reasons often requires argumentation. We need arguments as soon as a point of view, decision, or action needs to be justified, whether we are talking about introducing a new minimum wage or about whether Keanu Reeves is the worst actor in the history of cinema.[2] Arguments, then, serve a purpose that emotions, opinions, explanations, and descriptions cannot, but are not necessarily superior to other forms of engagement.

A GLOBAL DEBATE

Another factor to consider is that the world is getting smaller, in the sense that it is becoming increasingly likely that individuals—wherever they are in the world—will be asked to justify their decisions or actions, given that there is now so much more interaction among people who may previously never have met each other. If you run a South African company and want to set up an office in Singapore, the people you deal with in Singapore may have very different beliefs from you in terms of the ethics of business,

[2] Don't get us wrong, we love *The Matrix* and *Bill & Ted's Excellent Adventure*, but . . .

and they may also be subject to different laws than you are familiar with. Principles you might take for granted would need to be explained to them, and sometimes even argued for.

It is easy to understand why this is so. In the past, many individuals could live out their entire lives without being exposed to others who hold radically different opinions or beliefs. What this can lead to is a kind of complacence, in the sense that unchallenged opinion or belief very quickly gets treated as fact, and it becomes increasingly difficult to imagine the world as possibly being different from what you always imagined it to be.

In today's world, by contrast, we are constantly being exposed to different cultures and ways of life, and it is increasingly difficult to imagine that everything works the way you have always believed it to, or that your beliefs are certainly the correct ones to hold. Note that the person you are talking to might be equally convinced that his beliefs are right, even though they are opposite to yours. It would be unjustifiably arrogant for us to just insist that we are right and others are wrong, yet situations in which we might be tempted to do so occur more and more often as the impact of our global interconnectedness is felt.

So we need to develop resources to determine what is worth believing, or which actions are worth performing, without the safety net of our preconceived habits. A set of habits, cultural practices, or beliefs don't necessarily have to be bad or false—it is just that we should remember that they could originate from a sheltered view of the world and that those blinders are likely to handicap us when we deal with people who hold different views. Furthermore, we should remember that it is no longer possible to avoid this challenge, given that our interactions are increasingly going to involve people with an entirely different set of blinders.

We often neglect the importance of taking arguments and the reasons for our choices seriously. But if we never consider reasons why we could be *wrong*, we can never feel justified in believing that we are *right*—in other words, it is only through testing our beliefs, decisions, and actions in this space of reasons that we can know that we hold those positions responsibly, and that, in terms of our self-interest and the interests of others, they are the right positions to hold.

Careful reasoning, using the methodology of argumentation, is the most fruitful method for arriving at worthwhile beliefs as it recognizes the need for evidence, and is more careful and systematic than other methods. True, we might get lucky though deciding out of impulse, or through simply insisting that we are right because we'd rather not challenge our preconceptions. But if we are concerned with maximizing the number of times we actually do make optimal decisions, and minimizing the number of mistakes made, we should be concerned with understanding and using arguments when it is appropriate to do so.

CRITICAL THINKING—A SKEPTIC'S GUIDE

Let us now turn to some of the fundamental components of critical thinking. As you will see, even though critical thinking is a skill that you can never perfect, there are some rules of thumb that can help us become consistently better at reaching more optimal conclusions. These are:

- Extraordinary claims
- Falsifiability
- Occam's razor or parsimony
- Ruling out rival hypotheses
- Recognizing fallacies
- Separating induction from deduction

Extraordinary Claims

David Hume, an 18th-century Scottish philosopher, was an enormously influential force in the history of science and reason. In addition to inspiring Karl Popper's focus on inductive reasoning (described in Chapter 2 and in the following text), Hume was the first to outline a critical-thinking principle that was later expressed by Marcello Truzzi (1978) as "an extraordinary claim requires extraordinary proof," and then popularized by 20th-century skeptic Carl Sagan as "extraordinary claims require extraordinary evidence." (Sagan, 1980). Hume (1902) wrote that "A wise man . . . proportions his belief to the evidence." (p. 87). Later, in the same manuscript, he said that "No testimony is sufficient to establish a miracle, unless the testimony be of such a kind, that its falsehood would be more miraculous than the fact which it endeavors to establish." (p. 91)

These two guiding statements describe the critical thinker quite well. As discussed previously, a good skeptic or scientist believes a claim not unerringly, but instead only to the degree that it is supported by evidence. The second statement, on miracles, boils down to not accepting evidence for things that are quite unlikely unless the evidence is very strong in nature. Take this example:

> Your friend is late for a dinner you are hosting, and when he comes in his clothes are a mess with bloodstains and some type of hair. He explains that he had a car accident on the way, hitting a deer that was running across the road. It wasn't killed instantly, so he had to put it out of it's suffering, move it off the road, and then report the accident to the police and his insurance agency, hence his appearance and lateness.

Would you feel the need to doubt your friend's story? To have him take you to where the accident occurred and show you the deer carcass? Probably not, as you know that people sometimes have car accidents and that they are sometimes caused by deer running across the road. The claim is not very extraordinary, and the evidence for it (his wrecked car, the blood and hair on his clothes) fits the story well enough that most people would express sorrow over the accident, followed by happiness that your friend was unhurt.

But what if, instead of a deer, he told you that he had hit a Sasquatch or Bigfoot, the legendary man-ape that we will encounter in Chapter 9? Would you be likely to accept such a story on face value alone? Given its extraordinary nature, most of us would not, instead demanding to see the carcass of this legendary creature, or to have the blood and hair analyzed to determine whether they come from an unknown species. As Hume, Truzzi, and Sagan remind us: The more spectacular a claim is, the more solid the evidence must be in order to take it seriously.

Falsifiability

In Chapter 2, we introduced the idea of verification and falsification, emphasizing the importance of considering reasons why we could be wrong. Why should we try to establish whether our beliefs are false, rather than try to prove that they are true? This important distinction led to a revolution (credited mostly to Sir Karl Popper, who lived from 1902 to 1994) in the way science was conducted (Popper, 2002).

Prior to Popper's analysis, it was believed that we should attempt to prove the truth of our beliefs and scientific theories. But as Popper pointed out, with that model it was easily possible to demonstrate that very strange things were "true." For example, I can "prove" to you that it's through my mental powers that the sky does not fall on our heads. This claim is, after all, verified every day in that (a) I concentrate on keeping the sky up (it's a subconscious thing, so I can do it while I sleep too); and (b) the sky does, in fact, not fall on our heads.[3]

So if we use this model to justify our beliefs, it's just too easy to end up believing things that are actually false. Instead, he suggested that we rather think about *falsification* as the benchmark for what is worth believing. If you present a statement to me there must be some way of testing whether it is false—even if we aren't currently able to perform the test in question. The important point is that there must be conditions under which we would be compelled to say that the statement is false, because if there are no such

[3] Another example can be seen in a classic joke. A man is standing in Times Square in the middle of New York City. He is hitting two sticks together, over and over again. A policeman walks up to him, thinking he might be dangerous, and says "What are you doing?" The man replies, "I'm keeping the elephants away." The police officer says "Are you crazy, there aren't any elephants for thousands of miles!" The man smiles and says, "I know, I'm doing a good job, aren't I?"

conditions, the statement is effectively true by definition—it is presented as an axiom, rather than something that is responsive to evidence.

Of course, we would ideally be able to perform the test in question also, and discover that the statement in question passes the test. If a claim is falsifiable, and *also* survives our attempts to falsify it, we have good reason to consider the claim a strong one. With regard to the previous "sky falling on our heads" claim, it would be easy to test it by, for example, killing me. If the sky stays up, I was obviously talking nonsense.

Another consequence of Popper's work is that it seriously compromised various fields that were previously considered scientifically sensible or worthwhile, for example, astrology, where claims made are sufficiently vague or untestable that they are in effect immune to criticism. If I, as an astrologer, tell you that you're going to meet someone tall, dark, and handsome—and then you don't meet this person—it would be very easy for me to say that there must have been some gravitational influence from Planet X that caused my prediction to be false. There are so many complicating factors, I might patiently explain, that it's difficult to always get everything exactly right—but the last thing you should do is to question my methodology or the ancient and respectable art of astrology!

The point Popper made is that there is never any *possibility* of you being wrong, so long as you present your claims in a way that precludes them being tested, and that also precludes the possibility of their being able to fail any such test. This is what distinguishes science from pseudoscience. Besides the important implications it has had for science in general, it's worth pointing out one of the implications it has for us: If someone tells you something that can't possibly be falsified (we have no way of testing it, or it is designed to always be able to pass any test), then you have no rational reason for believing it. You may have emotional, spiritual, or some other sort of reasons, but you can't have rational ones. If you believe that reasons should ideally be rational, rather than rooted in bias, misconception, or superstition, then this insight has clear implications for what you should and should not believe.

Occam's Razor and Ruling Out Rival Hypotheses

The idea of falsification offers us a useful rule of thumb (or heuristic) for separating claims worth investing in from those that should be discarded (or at least, treated with less seriousness). Another device that serves a similar purpose is to consider various explanations (or hypotheses) for an observed phenomenon, and then to select the one that involves the fewest number of *additional* assumptions or other complications. This notion of choosing the most parsimonious explanation is commonly referred to as Occam's razor, after a Franciscan friar named William of Ockham (c. 1287–1347).

Although this heuristic cannot offer you a guarantee that you have selected the correct hypothesis, that is not its intent—instead, it relies on the idea that when you are trying to develop an explanation for an observation, you run a lower risk of error if you rely on assumptions that are not *themselves* in doubt or unprovable. For example, it's perfectly *possible* (in a logical sense) that you are the only person in existence, and that everyone else is a construct of your imagination. This notion in the philosophy of mind is known as *metaphysical solipsism* (Russell, 1988), and it takes the (less controversial) idea that we can only be aware of our own mental processes and not those of others to the rather more extravagant conclusion that therefore, we have no reason to believe that those other minds exist at all.

As noted previously, of course this is logically possible. But if you were to take metaphysical solipsism seriously, you'd have to offer an explanation for why the rest of us are so convinced of our existence, why we seem to share so many other people in common (in terms of our internal mental lives), and why you invented us—rather than more agreeable people—in the first instance. A far simpler explanation, and one that also conforms to our observations, is that it's instead true that we exist, and that we each have our own mental lives. Choosing the simpler (that is, fraught with less assumptions) explanation costs us no predictive power, and does not require us to assume something that seems to directly contradict our experiences (namely, that we exist). Using Occam's razor to sort through rival hypotheses can allow you to winnow the field of potential explanations down to only those that are reasonable quite quickly. Those that are left can then form the basis for testing through scientific methodologies, in order to see which has the most support.

Fallacies

Fallacies (or fallacious reasoning) are very common—if you pick up any newspaper or magazine, you're likely to find at least one example in its pages. Fallacies are even more common in day-to-day conversation, partly because they help to make some of the stories we tell ourselves about evidence, and what it means, more plausible. When we generalize from our own experience to a general rule, for example, we are reasoning fallaciously, in that we are forgetting that our individual experience might be an exception that can't reliably be generalized.

A fallacy is an error or mistake in reasoning, but one that takes a certain defined form. There are some sorts of error that are so common that we have names for them, for example, the Latin *ad hominem*. This phrase, literally meaning "against the man," describes the tactic you might know as "playing the man, not the ball." Just in case you're not familiar with this tactic, it refers to attacking a person as a way of avoiding dealing with her arguments. For example, one might say "She's too young to know what she's talking about," and in dismissing the speaker out of hand in this fashion, never notice that she was saying some very important things.

As mentioned previously, one very common way of making a mistake in reasoning is to imagine that your own personal experience is true for everyone—to generalize from a sample of one (you) to a sample of all (the world), in other words. For example, if you have a bad experience at a restaurant, you're quite likely to recommend to your friends that they eat elsewhere. But this recommendation would be based on the fallacy of *hasty generalization*—in which we generalize from a set of observations that is too small or unrepresentative to justify the point we're trying to make.

Now, the fact that you've reasoned fallaciously does not mean that you're *wrong* that it's a bad restaurant: the fact that a conclusion is reached through poor reasoning does not by itself establish that the conclusion is false. It could be a true conclusion, even if your reasoning is bad. But neither you—nor anyone else—have a *justified reason* to think it a true conclusion (or a false conclusion, for that matter) until a sound argument settles the matter.

An exhaustive catalogue of the various logical fallacies could take up many chapters of this book—there are many fallacies one could identify, and there are also many disagreements one could identify with regard to those fallacies. For example, it's not always immediately obvious whether or not the identification of a detail about a person (rather than a focus on his argument) is fallacious or not. If you think that part of a political leader's job is to manifest a certain sort of moral virtue, then you would think it entirely relevant to his suitability for the job that he was engaged in an extramarital affair. By contrast, if you think that a political leader's job is to spend taxpayers' money wisely, and administer the affairs of state in a fair and efficient manner, you'd perhaps think that detail less relevant or even irrelevant.

We'll thus defer to texts that are exclusively about informal logic, including fallacies, for discussion of these complexities as well as a comprehensive list of the various fallacies you might encounter. There are also numerous Internet-based resources you could consult—the Nizkor Project (n.d.) is well regarded, as is the site Your Logical Fallacy Is (n.d.); the websites for both can be found in the references for this chapter. The other reason why this text won't focus on such a list is that an obsessive focus on fallacies can itself conduce to poor reasoning, as we'll explain now.

Engaging With Fallacies

After learning about fallacies for the first time, it can be entertaining to find them everywhere, and to enjoy pointing out people's errors to them. But that is often not a very useful approach, especially if you limit yourself to snobbish references to the Latin word describing the fallacy, without bothering to explain what is wrong with the argument. This isn't useful because it minimizes the possibility of learning—the person who has committed the fallacy will learn nothing from you just mentioning the name of a fallacy (unless they happen to know the fallacy already). You've also made it less likely that you can learn from the conversation, because you haven't offered much opportunity for any debate that might expose errors in your own thinking. So, the names of the fallacies are simply tools to help us identify and remember them—they are not meant to be weapons that you use as a way to shortcut any reasonable debate. It's also far less important that we know the names (whether in English or in Latin) than that we can explain how the fallacy is an example of poor reasoning.

Second, we learn about fallacies not only to detect them in the arguments of others, whether in writing or in conversation, but also to detect and avoid them in our own thinking. Critical thinkers are concerned with improving their own arguments and making more sense themselves, not only with being critical of the arguments of others. A mistake pointed out earlier in the chapter—that of thinking that because a person has reasoned fallaciously her conclusion must be false—is important enough to highlight in concluding this section. The *fallacy* fallacy, an error of relevance, points to the fact that whether or not *particular elements* of reasoning are poor is not necessarily *relevant* to the truth of a conclusion. One can make all sorts of mistakes in reasoning, and yet—by good fortune, or because of other redeeming features of the argument—end up with a true conclusion. Likewise, one can reason fairly well, and end up with a false conclusion. This is why we'd urge you to remember that fallacies are merely a heuristic, or shortcut, to identifying flaws in reasoning. They are not the end of an argument, and the argument in question might be able to proceed entirely productively, even once one removes the fallacious component it contains.

Induction Versus Deduction

Chapter 2 offered an outline of the difference between *deduction* and *induction*. You'll recall that induction involves making an inference about a rule or pattern by reference to specific observations. These inferences can be so well justified that they might as well be treated as fact—for example, our (limited) observations that the sun has always risen in the East[4] gives good grounds for thinking that the rule "The sun rises in the East" is true and that your *prediction* that the sun will rise in the East tomorrow will also end up being true. As quoted in Ellis (1970, p. 431), the Danish physicist Niels Bohr is said to have quipped "prediction is very difficult, especially about the future"—but we can make it the case that our predictions are less or more reliable, depending on how many

[4] Unless, of course, you have happened to be very close to the North or South Poles, but, let's be honest, you haven't been.

observations we base the prediction on, and whether those observations are reliable and relevant to the prediction we're making.

By contrast, deductive arguments can quite reliably generate conclusions, but at the expense of being more modest in their ambitions. As Chapter 2 noted, deductive arguments tend to progress *from* general rules *to* specific claims, rather than using specific claims to generate rules or predictions. For example, this classic Aristotelian syllogism (a syllogism is typically a three-line argument, with a major and a minor premise, followed by a conclusion) generates a conclusion that you cannot reasonably doubt, while simultaneously being thoroughly unimpressed by its lack of ambition:

1. All men are mortal.
2. Socrates is a man.
3. Therefore, Socrates is mortal.

As you can see in this example, the conclusion is *narrower* in scope than the major premise ("all men are mortal") in only referring to one man, Socrates, whereas a claim that the sun will rise in the East tomorrow is *broader* in scope, in that it makes a prediction about the future. This is an essential difference between inductive and deductive arguments, and is perhaps best explained by reference to the capacity each type of argument has to preserve the truth of its premises.

Logic and Truth Preservation

Logic can be very useful for preserving truth. What this means is that if you have true evidence, that evidence will usually result in true conclusions *if your reasoning is free of error*. And seeing as fallacies are one type of error, avoiding fallacies is a good way to maximize the chances of reaching true conclusions. But logic does not guarantee that the conclusions you reach are certain to be true. Consider this example:

1. Grass is red.
2. All red things are poisonous.
3. Therefore, grass is poisonous.

What's interesting about this example is that the (deductive) argument presented is in fact *valid*. It's a valid argument because the conclusion follows necessarily from those pieces of evidence (premises). To put it another way, *if* (or assuming that) the premises are true, the conclusion *must* be true also. What this shows is that logic can be quite technical, or removed from reality, because this is obviously not a "good" argument in other ways.

But it's important that logic is technical, because we sometimes don't understand reality as well as we should, and we sometimes don't have the evidence we need to evaluate something. So, logic, or the structure of arguments more generally, gives us a second sort of way to engage with arguments. We can either see whether the evidence is good (that's the first way), or we can see whether the argument is constructed in a strong, reliable sort of fashion (that's the second way, and involves considering the principles of logic).

An inductive argument can never be *valid*, at least if you use the word in its technically correct sense. Even the example referred to earlier, regarding the sun rising in the East, is not valid. Let's say the argument looks like this:

1. The sun has risen in the East for all of recorded history.
2. We have no reason to believe that tomorrow will be any different than those millennia of recorded history.
3. Therefore, the sun will rise in the East tomorrow.

Now, only a fool would bet against the sun rising in the East tomorrow. However, the fact that it will do so is not *guaranteed* in a logical sense—some cataclysmic cosmological event might mean that it does not (of course, we'd no doubt end up dead if that were the case). The point is not whether this outcome is likely or not, but rather that validity deals with a cast-iron guarantee that is offered without reference to probability or prediction, but that is instead offered by virtue of an argument's *structure*.

For that reason, deductive arguments have the virtue of being able to preserve truth (*if* the premises are true, and the argument is valid, *then* the conclusion would also be true), but at the expense of not being knowledge expanding (e.g., the conclusion cannot make inferences about the future). Inductive arguments can be knowledge expanding, but at the cost of not being able to guarantee truth preservation.

Having said that, perhaps truth isn't always attainable, at least if we understand it to mean "certainty." How often are we able to be certain about anything? Take, for instance, a very well-justified claim in science—not as well justified as the fact that the sun will rise in the East, but still one that you'd be a fool to doubt—namely, that smoking tobacco products contributes to developing cancer. We can be confident that this is true, without being certain of it. This is because more important than certainty is how well justified a conclusion is—and when it's sufficiently well justified, it's reasonable to regard it (and speak of it) as true.

CONCLUSIONS

Being a good critical thinker isn't always easy, and as you'll see in the next two chapters, even the best critical thinkers need to deal with the confounds of how our brains (Chapter 5) and our worlds (Chapter 6) get in the way of our ability to reach the most rational decisions. However, understanding the principles of critical thinking is an essential foundation for making rational decisions, and the basic principles are easy enough to remember and implement when possible. But an important thing to remember is that if you focus on certainty—on having some sort of guarantee that you're making an optimal choice, or behaving perfectly rationally, you're likely to end up being unable to meet that standard most of the time. It is rarely useful to insist on knowing, for sure, that something is true—simply because it's very seldom the case that we can *know* something to be true, with absolute certainty, and by that standard most of our conversations and arguments would be pointless.

Instead, the principles described in this chapter (see Figure 4.1 for a reminder) help us to assess when a belief is *justified*, rather than when it is *certain* to be true. Asking whether a belief is justified means asking yourself a hard question: Does the available evidence, when evaluated fairly, make a certain belief more likely to be true than any competing beliefs?

FIGURE 4.1 Critical-thinking principles.

Arguments serve the function of offering justification for their conclusions—that's their job, to put it simply. A good argument will give us good reason to accept the conclusion as being true—even if only provisionally, seeing as new evidence may come along and change our minds. But if you encounter a good argument for a conclusion that you don't want to accept, you're stuck with a problem—either you need to change your mind, and accept the conclusion, or you need to show that the argument is not a good argument, which may require you to offer a counter argument of your own, showing that the original argument is not in fact a good one. The option you don't have (or rather, shouldn't take) is to keep believing in the conclusion you prefer, even though the arguments supporting it are weak.

So, arguments offer justification for their conclusions. But arguments contain premises, or evidence, and sometimes we also need guidance on when to consider premises justified. Remember, we don't need to know whether they are true—we just need to have good reason to believe that they are acceptable, in that they are justified, and therefore likely to be true. The question to ask yourself about a premise is: Is it reasonable to accept this as being true? If so, we can consider whether the reasoning—the way in which the premises link to the conclusion—is sound.

If it is reasonable to accept the premises, and the reasoning is sound, the conclusion is also more likely to be true. There are no guarantees that you'll always get things right, but careful and objective assessment of the evidence and arguments is our best resource for avoiding the most egregious errors to which we might otherwise fall prey. That careful and objective analysis is often complicated by how we process information, as we'll see in the next chapter, which describes why we can't trust our brains.

QUESTIONS FOR REFLECTION

1. *In terms of Chapter 1's discussion on knowledge and justification, how would you describe the main problems with the views held by those we describe as "hyperskeptics?"*
2. *When engaged in argument, is it ever a good idea to "agree to disagree?" What are some examples? How would you identify those situations?*
3. *How has globalization impacted the arguments that you encounter in your day-to-day life?*
4. *If you audit your own beliefs thoroughly, do you think that you hold any unfalsifiable beliefs? Do you think you ought to discard those beliefs? Why or why not?*
5. *Imagine that a friend reported what seems to be a very improbable experience to you—perhaps that he was abducted by aliens. How would you go about trying to make your friend think critically about that experience?*

REFERENCES

Briggs, R. (2014). Normative theories of rational choice: Expected utility. In E. N. Zalta (Ed.), *The Stanford encyclopedia of philosophy* (Fall 2014 ed.). Stanford, CA: Stanford University. Retrieved from http://plato.stanford.edu/archives/fall2014/entries/rationality-normative-utility

Butler, H. A. (2012). Halpern critical thinking assessment predicts real-world outcomes of critical thinking. *Applied Cognitive Psychology, 26*(5), 721–729.

Sagan, C. (writer/host) (1980, December 14). *Encyclopaedia Galactica. Cosmos.* Episode 12. 01:24 minutes. Public Broadcasting Service.

Committee for Skeptical Inquiry. (2014). *Deniers are not skeptics.* Retrieved from http://www.csicop.org/news/show/deniers_are_not_skeptics

Edwards, K., & Smith, E. (1996). A disconfirmation bias in the evaluation of arguments. *Journal of Personality and Social Psychology, 71*(1), 5–24.

Ellis, A. (1970). *Teaching and learning elementary social studies.* London, UK: Pearson.

Hume, D. (1902). *An enquiry concerning human understanding.* Retrieved from https://www.gutenberg.org/files/9662/9662-h/9662-h.htm

Loxton, D. (2013). Why is there a skeptical movement? Retrieved from http://www.skeptic.com/downloads/Why-Is-There-a-Skeptical-Movement.pdf

Nizkor Project. (n.d.). *A comprehensive list of fallacies.* Retrieved from http://www.nizkor.org/features/fallacies

Popper, K. (2002). *Conjectures and refutations: The growth of scientific knowledge.* London, UK: Routledge.

Russell, B. (1988). *The problems of philosophy.* Buffalo, NY: Prometheus Books.

Sellars, W. (1956). Empiricism and the philosophy of mind. In H. Feigl & M. Scriven (Eds.), *Minnesota studies in the philosophy of science* (Vol. I, pp. 253–329). Minneapolis, MN: University of Minnesota Press.

Simon, H. (1978, December). *Rational decision-making in business organizations.* Nobel memorial lecture. Retrieved from http://www.nobelprize.org/nobel_prizes/economic-sciences/laureates/1978/simon-lecture.pdf

Truzzi, M. (1978). On the extraordinary: An attempt at clarification. *Zetetic Scholar, 1*(1), 11.

Your Logical Fallacy Is . . . (n.d.). *Thou shalt not commit logical fallacies.* Retrieved from https://yourlogicalfallacyis.com

CHAPTER 5

WHY CAN'T WE TRUST OUR BRAINS?

The human brain is immensely complex, with an estimated 86 billion neurons and some 100 trillion synapses (Azevedo et al., 2009). It is amazingly resilient in the face of injury, able to reroute functions around damaged or purposefully removed areas to restore partial or full functioning. It can store an almost limitless amount of information for decades when healthy. However, even with these wondrous properties, the brain is quite easily fooled. This means that you need to seriously doubt . . . yourself!

A few years ago, when several students and I (CWL) started a skeptical/freethought campus group, the Center for Inquiry sent a giant box of goodies. In it were materials to help the group in getting started: flyers, copies of *Skeptical Inquirer* and *Free Inquiry*, stickers, and more. One of the stickers was a play on the Tide detergent logo and slogan that said "DOUBT—For even your strongest beliefs!" I loved it (and, in full disclosure, kept that one and put it in my office, where it is to this day).

Doubting is one of the hallmarks of an enlightened mind and key to being a good scientist and a good skeptic. It is not enough to only doubt others, which we all do. Instead, people must also be willing and able to doubt their own beliefs and convictions across all areas of life. Many people do not do this though, and instead plow through their lives convinced that their beliefs and perceptions of the world are accurate and unimpaired. The purpose of this chapter is to teach you some of the many, many reasons why you cannot blindly trust your own brain and to prove that doubt (even of yourself) is not just a good thing, but a necessary thing.

Many researchers in the psychological and other sciences have spent decades looking into specific ways that the brain can be fooled. This chapter first focuses on two broad types of ways that our brains predictably prevent us from making logical, rational decisions. First, we discuss cognitive biases and mental heuristics that we all have, and how those often cause us to perceive a world that conforms to our preconceived notions about how things *should be*, rather than how they are. Next, we will look at how humans naturally misperceive and misevaluate the data we are exposed to, especially in the case of ambiguous information. The chapter concludes by examining why we have these specific types of problems with data. In particular, we consider the concept of bounded rationality, which describes how our pattern-recognition abilities and motivation to find reasons for events that occur in the world around us yields multiple benefits for our species, evolutionarily speaking, while also resulting in particular problems with evaluating information. We also examine how using mental shortcuts can help us be more cognitively efficient, but again results in bias when presented with new or inconsistent information.

THE LOGICALLY ILLOGICAL BRAIN

Thanks to the hard work of hundreds of researchers over the past half century, we have developed an ever-increasing understanding of the myriad ways that humans do not act rationally or make optimal decisions even when presented with straightforward information and data. Instead, we often act and think in an understandable but irrational manner—what we are calling "logically illogical." Exploring the various ways in which we can make poor decisions can then, in turn, lead us to understand why doubt of yourself is crucial to becoming a critical thinker.

Two of the largest factors influencing why you should doubt yourself frequently and thoroughly are *cognitive biases* and *mental heuristics*. Since Kahneman and Tversky's (1972) landmark article on how we, as humans, make nonrational decisions, the research on these factors has grown immensely. Cognitive biases are predictable patterns of judgmental deviation that occur in specific situations, which can cause inaccurate interpretation or perception of information. In other words, cognitive biases are regularly occurring ways that our brain misinterprets evidence. They impact our ability to make accurate, logical, evidence-based decisions on a consistent basis. Recognizing cognitive biases helps us to realize that our intuitive understanding of the world is often (but not always) distorted in some way. Once you understand that you cannot always trust your own judgment, and therefore need to apply doubt to yourself, you are on your way to beginning to overcome some of them. Luckily for us, these are "predictably irrational" (Ariely, 2009) problems, and so, by becoming aware of them on a conscious level, we are able to combat their influence.

Slightly different, heuristics are mental shortcuts or rules of thumb that significantly decrease the mental effort required to solve problems or make decisions (Kahneman, Slovic, & Tversky, 1982). Unfortunately, heuristics often lead to an oversimplification of reality that can cause us to make systematic errors that can then become cognitive biases. It is critical to note that just because we make the mistakes outlined in this chapter (and boy, do we make them), this does not mean that humans are terrible decision makers all the time. In fact, in many situations it is actually adaptive for us to use heuristics and mental shortcuts (something we come back to at the end of the chapter). Indeed, ignoring information can actually improve one's judgment in specific situations. In their review of a massive amount of literature, Gigerenzer and Gaissmaier (2011) note the main impacts of heuristics on decision making.

First is a positive benefit called the "less-is-more" effect. Sometimes heuristics result in as or even more accurate results than more involved strategies, despite processing less information. Second is what's called *ecological rationality*. This means that heuristics aren't inherently good or bad, but that their accuracy depends on the environment they are used in. Over time, people (generally) learn which heuristics are most useful to them in their particular environment. It is important to note that real-world decision making frequently uses heuristics not because they are necessarily the best way to make decisions, but because we can't access all the needed information to make a perfect decision. Heuristics allow us to take a mental shortcut and arrive at a "good enough" decision.

Although there is not a "standard list" of cognitive biases and heuristics (and, in fact, quite literally hundreds have been identified), the ones described here are some of the most well-researched and common biases/heuristics we encounter as humans:

- Confirmation bias
- Belief perseverance
- Hindsight bias
- Representativeness heuristic
- Availability heuristic
- Anchoring and adjustment heuristics

Confirmation Bias

Confirmation biases are one of the most frequently encountered, most frustrating, and yet most understandable biases (Nickerson, 1998). Confirmation bias is the tendency of individuals to favor information that confirms their beliefs or ideas and discount those that do not. This means that, when confronted with new information, we tend to do one of two things:

1. If this information confirms what we already believe, we throw it a huge "welcome home!" party. We unreservedly accept it, and are happy to have been shown it. Even if it has some problems, we forgive and forget those and welcome this new information into our brains with great

fanfare. We are also more likely to recall this information later, to help buttress our belief.

2. If this information contradicts what we already believe, we slam the door in its face and tell it to get the hell off of our porch and never come back or we will call the police. We nitpick any possible flaw in the information, even though the same flaw would not get a mention if the information confirmed our beliefs. It also fades quickly from our minds, so that in the future we cannot even recall being exposed to it.

Here's an example. Let's say that you, for whatever reason, think that the breed of dog called pit bulls are dangerous animals. When a friend tells you about an accident in which a dog attacked a child, and mentions that it was a pit bull, your brain will glom onto that, and you will say to yourself (and maybe your friend), "See, I knew they were dangerous, terrible beasts!" However, if the same friend later tells you she is dating someone who owns a pit bull, and how well behaved, sweet, and affectionate it is, you are likely to discount that information in some way ("Oh, she's just in love and isn't objective" or "That dog is an exception to the rule").

The confirmation bias could even be seen as the entire reason that the formal scientific method (as discussed in Chapter 2) had to be developed. We naturally try to find support for and prove our beliefs, which can in turn lead to the wholesale discounting or ignoring of contradictory evidence. Science, in contrast, actively tries to *disprove* ideas. The scientific method allows for increased confidence in our findings and makes scientists less prone to the confirmation bias.

If you have ever been in an argument about an issue that you care deeply about, chances are you have experienced an interesting aspect of the confirmation bias: the more emotionally charged or deeply held our beliefs, the stronger the effect of the confirmation bias (Plous, 1993). This aspect of the confirmation bias underlies one of the primary reasons why reasoning and producing facts does not work very well in most debates and arguments: We have already made up our minds and ignore that information which shows us to be wrong.

Belief Perseverance

Have you ever told someone that something she believed was demonstrably wrong . . . only to later have her still express belief in it? Or, perhaps you have clung too tightly to beliefs and ideas even when you should have changed? If so, you are already familiar with the effects of belief perseverance. Stated simply, belief perseverance is the tendency to stick with an initial belief, even after receiving contradictory or disconfirming information about that belief (Anderson, 2007).

Speaking broadly, we as humans tend to show three kinds of belief perseverance: self-impressions, social impressions, and naïve theories. You may believe you have a beautiful singing voice, despite your family evacuating the house when you take a shower (self-impression). Or perhaps you dislike a coworker, even after he has done several nice things for you (social impression). Maybe you have a specific belief about how the world works, for example, that the Earth is stationary and the sun revolves around it, and even when presented with the overwhelming scientific evidence showing the heliocentric nature of the solar system you go with how you think things work (naïve theories).

Our beliefs stay believed for a number of reasons, among them the availability heuristic (see the section later in this chapter for details on that one), illusory correlations, and distortions of evidence. The illusory correlation refers to seeing a relationship between two things (events, people, places, activities) when in reality none exists (Eder, Fiedler, & Hamm-Eder, 2011). Superstitions are prime examples: you see a black cat moving across the street, cross its path, and then shortly thereafter get a flat tire that makes you late for your job. You then look back and think, "I knew I should have gone down a different road!" In reality, of course, unless the cat had ninja-like skills and was able to actually jump onto your car and slice the tire, the cat had nothing to do with it. A more real-world example would be praying for someone you care about to get better in midst of an illness. Lo and behold, the person does get better, which you then attribute to your intercessory prayer. In reality, carefully controlled studies and meta-analyses have found no link between health outcomes and prayer (Benson et al., 2006; Masters, Spielmans, & Goodson, 2006).

Distortions of evidence can be explained, in part, by confirmation bias. To continue the preceding example, because you believe that intercessory prayer is effective, you may remember those times when you prayed and someone got better, but ignore the times your prayers were not or (more often) discount those times as being "beyond my understanding." In short, you remember the hits, forget the misses, and distort the actual data in favor of making it support your beliefs.

Hindsight Bias

If you have ever told someone that you knew something was going to happen, after it already did, you have likely fallen prey to the hindsight bias. The hindsight bias occurs when we overestimate how confident we are in an outcome after the outcome is already known (Roese & Vohs, 2012). This bias is commonly described as the "I knew it all along" fallacy.

An example of hindsight bias would be hearing about a dating couple breaking up and saying, "Oh, I absolutely knew that they were not right for each other all along." Chances are, before you knew the outcome, you would not have used the term "absolutely" (unless they were just a horrible mismatch). You may have said they "probably" wouldn't make it, or that there "wasn't much of a chance" when they were dating, but once you knew the result, you became much more certain of your previous estimation of the outcome.

Rather than a single construct, there appear to be three distinct forms of hindsight bias: memory distortion, inevitability, and foreseeability (Blank, Nestler, von Collani, & Fischer, 2008). In memory distortion, we inaccurately recall our earlier estimate of something occurring, which in turn distorts our current estimate ("I said it would happen"). The inevitability form of bias occurs when one believes a past event was predetermined for some reason ("It had to happen"). The last kind of hindsight bias is the foreseeability form. This happens when you think that you could have foreseen something that has already occurred as occurring ("I knew it would happen"). Each of us is prone to each of these kinds of bias, but there is interesting work that shows different types of inputs (or reasons) can impact what type of hindsight bias we show (for a review see Roese & Vhos, 2012).

Representative Heuristic

At some point in our past, each and every one of us has, despite admonitions against doing so, "judged a book by its cover." You see someone (or something), and she, he, or it appears to belong to a particular category, so you attribute certain qualities to her, him, or it (Kahneman & Tversky, 1972). Sometimes you are terribly wrong. This heuristic can be thought of as a form of stereotyping, of taking a particularly salient feature of someone or something and overgeneralizing it inappropriately.

For example, let's say that your authors, as teachers at a university, see two students in their class. One is well groomed and well dressed, sits in the front row, appears to be paying close attention to the lecture, and is taking copious notes. The other student is dressed in sweatpants and an old T-shirt, is playing on her phone, and appears oblivious to anything happening in the class. Our initial reaction could be to say that the first will do well in class and the second is likely to do poorly. In doing so, we are comparing their outward appearances and behaviors to our idea of what a "good" student looks like. In other words, one of them is representative of what we think a good student looks like and one is not. Yet in reality it could be that the student who seems to be on top of things is completely clueless whereas the other is bored because she has read ahead and studied the topic on her own and wants to get deeper into the issue. Of course, we could also be right in our judgment. Only time (and each one's grades in the class) will tell!

There are several reasons why we misjudge people in this way. First, we ignore the base rates of behaviors, or how common something is in a given population. For instance, your friend tells you about a person she knows who is shy, good at mathematics, and loves Star Trek. She then asks you to guess what his major was in college—business or mechanical engineering? Many people will guess he is an engineer based on the descriptors, and completely ignore that (statistically speaking) he was much more likely to be a business major, as there are many more business than engineering majors on most college campuses.

Another reason why the representative heuristic occurs is that we often draw inappropriate conclusions from small sets of data. If you have only met two people from the United Kingdom, and they are similar, you may think that they are representative of all the people from the United Kingdom (when they could, in fact, be the only two people like that in the entire British Isles). This process can be extended to any particular group of people (e.g., race, profession, geography, religion) or objects (e.g., model of vehicle, brand of clothing, style of furniture).

Availability Heuristic

When making decisions, we tend to be biased by information that is easier to recall; such information could be that which is more vivid, well publicized, or recent. Such easily retrievable information can cause the availability heuristic to occur, when we make judgments about how likely something is to occur based only on how easily it is brought to mind (Schwarz et al., 1991). This can cause a number of problems in how we process information and make choices.

Much of the research on the availability heuristic has used a protocol similar to the following (adapted from Combs & Slovic, 1979). First, subjects are presented with a question ("Which of the following causes more deaths in the United States per year?"). They then have to make a choice between two options in response to this question, one of which is much more easily recalled because of news coverage or recency to an individual (e.g., lung cancer vs. automobile accidents). Even though lung cancer kills three or more times as many people as car crashes each year, a majority of people tend to choose car crashes as being more prevalent. This choice could be made because they can easily recall a crash that they witnessed, were a part of, or saw on television or the news, but have a harder time retrieving information on a lung cancer death (unless, of course, they have had recent exposure to a person or story involving such a death).

For many readers of this book, the impact of recency may be particularly salient in activating the availability heuristic. Imagine that you are engaged in a debate, online or in real life, with someone who holds non–evidence-based beliefs on some topic, like the idea that childhood vaccines cause autism. The person may be rude, antagonistic, mean, or just unpleasant in some way. You leave the conversation frustrated, and almost immediately encounter someone else who reveals, at the start of your interaction, that he or she also thinks vaccines cause autism. You are very likely to attribute to the new person the characteristics of the old—that he is rude, mean, and so on—even without any evidence to back that up.

Anchoring and Adjustment Heuristics

When we try to estimate various phenomena (population of cities, number of murders committed per year, percentage of college students who graduate in 4 years) that we know little about, our answers can be quite easily manipulated (Epley & Gilovich, 2006). For instance, if you are asked, "Is the percentage of bald Americans more or less than 15%?" instead of, "What is the percentage of bald Americans?" you are likely to give very different answers. This problem is called the anchoring and adjustment heuristic, and demonstrates how our estimates are influenced by initial anchors. It can occur even when given anchors that are obviously ridiculous (Strack & Mussweiller, 1997), as in the following example.

Suppose someone asks you, "Is the percentage of Americans who are skeptics greater or less than 0.0005%?" and then later asks you to give an estimate of the number of Americans who are scientific skeptics. This person then asks your best friend, "Is the percentage of Americans who are skeptics greater or less than 95%?" and later asks him to give an estimate of the number of Americans who are scientific skeptics. Odds are, you will give a much lower number than your friend will, even if you may both have given similar estimates if I had simply asked, "What percentage of Americans are skeptics?"

Why does this happen? It appears that our estimates stay too close to the original anchor, regardless of its absurdity, and we fail to take into account other sources of information that may help to provide a more reasonable answer. Our final estimates are off by more than they would be *sans* anchor because we stop once we reach the edge of a plausible range for the estimate. It is likely that the true estimate is closer to the middle of the plausible range but, thanks to the anchor, you stay closer to the side of your anchor.

GETTING SOMETHING FROM NOTHING

Have you ever seen or heard something that wasn't actually there? We aren't talking about having a hallucination (but if you have had some, you aren't alone—they are surprisingly common; Sacks, 2012), but instead about misperceiving information in the world around you. Perhaps you were walking outside, looked up into the sky, and saw a cloud that looked like an elephant. Or you went to eat your toast only to discover the unexpected image of a man's face staring back at you. Or maybe you've been singing a new favorite song for months, only to discover that the actual lyrics are not what you've been hearing all this time.[1] We frequently see things or hear things that aren't actually there, because that is what we are primed to see or hear, in large part because of our confirmation biases. Let's take a look at some common examples first, then discuss why this occurs.

Backmasking

In the 1970s and 1980s in the United States, there was a minor uproar among concerned Christian parents about Satanic and occult messages that were being spread by rock and roll bands like Led Zeppelin, the Rolling Stones, Queen, the Eagles, and Black Sabbath. They weren't spreading these messages overtly, though. Instead, these bands were stealthily sneaking them in via *backmasking*—a technique in which a sound or message is recorded backward onto a track that is meant to be played forward. The idea was that the messages (which allegedly said things like "Here's to my sweet Satan" and "It's fun to smoke marijuana") were going to subliminally influence listeners to worship the devil, have sex, and use drugs. This was such a concern that a hearing before the U.S. Congress was convened on the issue and lawsuits were even brought against the heavy metal band Judas Priest after two young men committed suicide, purportedly in response to a backmasked message in their music.[2]

The Face of a Martian

In 1976, the first spacecraft to reach Mars began sending back images of the planet's surface. The Viking I Orbiter sent back loads of data with incalculable scientific value, but perhaps the most famous image was the so-called "Face on Mars." Figure 5.1 shows the original photo, with the area in question enlarged. The image depicted what appeared to be a humanoid face approximately 2.5 kilometers long, with recognizable eyes, mouth, and a nose. Dismissed by the National Aeronautics and Space Administration (NASA) scientists working on the project, the Face nonetheless was claimed by others to be the work of

---

[1] When one of your authors (CWL) was younger, he thought the classic song "*I Can* See Clearly Now" had the very confusing line "I can see clearly now, the rain has gone . . . I can see all popsicles in my way . . . ." The actual word is "obstacles," of course, but as a child he marveled at this mysterious world of grown-ups, where popsicles were so numerous that one had to go out of one's way to avoid them.

[2] At this point, we highly recommend listening to some of these supposed backmasked messages. Canadian educator Jeff Milner has a large collection of the clips on his website (jeffmilner.com/backmasking/), allowing you to play the songs forward, and then in reverse, and then reveal what the backmasked message "is."

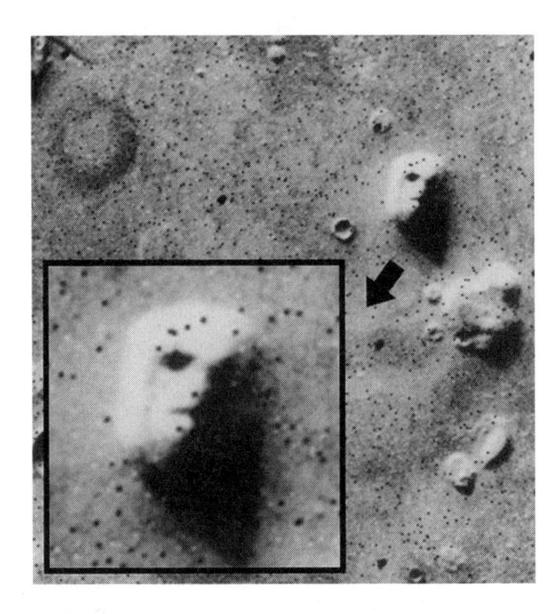

FIGURE 5.1 Low-resolution "Face on Mars"
Source: Courtesy of National Aeronautics and Space Administration (NASA). Image captured by Viking Orbiter 1, July 25, 1976.

a sentient species, an alien race. Several proponents of this idea also claim to see evidence of purposefully constructed pyramids and other structures in the Cydonia region of Mars where the Face was found.

Religious Imagery in Unusual Places

In the spring of 2005, Obdulia Delgado was driving in Chicago on her way to a final exam. While going through an underpass, something caught her eye on the wall. She screeched to a halt, disrupting the busy traffic around her. Her eyes were transfixed by the image of a woman's face, surrounded by a cloak. It was an image of the Virgin Mary. She later recounted that "I was so stunned I couldn't move. People were honking. It was a dream. I don't even know how I got home" (Lebovich, 2005). Within a short period of time, crowds of people had gathered, lighting candles, saying prayers, and making a memorial around what the Illinois Department of Transportation described as a stain caused by saltwater runoff.

This event is far from unusual, as people have reported images of Mary in windows, toast, pizza pans, and firewood. Not to be outdone, images of Jesus have been seen in trees, tortillas, clouds, and even in the deep space Cone Nebula (as shown in Figure 5.2, an image taken by NASA's Hubble Space Telescope). Muslims often report seeing the Arabic word "Allah" in places as diverse as eggs, plants, fish, and even on a sheep. Western religions do not have a monopoly on this, either. In Singapore there is a very famous

FIGURE 5.2 Cone nebula
Source: Courtesy of National Aeronautics and Space Administration (NASA). Image captured by the Hubble Space Telescope, April, 2002.

"Monkey Tree" that supposedly bears the image of two or three monkeys in its bark; both Hindu and Chinese people have prayed at the site and set up shrines to monkey-related deities from their respective religions.

Sports Curses and Jinxes

In the United States, one of the most popular videogame franchises is the *Madden NFL* series, an American football simulator. Games have been released annually for over 25 years and, starting with *Madden NFL 99*, feature a very prominent player from the prior year's season on the cover. Although many would consider such an opportunity as fantastic, others have seen it as a curse. The so-called "Madden Curse" refers to how almost all of the cover stars of the videogame have, in their next season, played very poorly or suffered debilitating injuries. One article on the Madden Curse reports that all but one player became cursed after being on the cover (Marshall, 2014). This has even led some players to reportedly turn down the offer to appear on the cover (and the very large paycheck that would undoubtedly go with it).

Following a similar vein is the "*Sports Illustrated* jinx," when those athletes or teams that grace the cover of the weekly sports magazine reportedly have bad luck or do poorly soon afterward. This dates back to the initial cover in 1954, which featured a baseball player who suffered a hand injury immediately after the issue came out. On their website, *Sports Illustrated* even offers a list of all the reported jinxes since 2005, a total of 21 problems that impacted basketball, baseball, football, and soccer players.

Seeing Because of Believing

All of us occasionally mishear or misunderstand things; this comes as no surprise to anyone who has ever interacted with another human being. However, the human brain appears especially good at misperceiving information in a particularly biased fashion, especially if that information is uncertain or unclear. All the previous examples (Satanic lyrics in rock songs played backward, religious imagery in mundane objects, magical curses on athletes) show how readily and easily the brain engages in *pareidolia*, the tendency to interpret random or ambiguous stimuli as being something clear and distinct (Zusne & Jones, 1989). As several of these examples illustrate, we tend to experience pareidolia for human faces and religious-based images in particular. Understanding why requires unpacking two interesting aspects of human psychology—priming and patternicity.

Priming is an implicit, nonconscious form of memory in which being exposed to stimuli in one context changes your behavior in another context, often making you more likely to recognize, recall, or otherwise pay attention to similar stimuli (Marsolek, 2003). For example, if I give you a list of words to look at that includes the word "king," and then later ask you to complete the word "K I _ _," you are more likely to spell "king," than "kill" or "kind," because you have been primed. One clever study (Harris, Bargh, & Brownell, 2009) had children watch a cartoon and eat a snack. Half the children saw advertisements for food while watching, the other half saw advertisements for toys and games. The primed children (who had seen advertisements with people eating and talking about being hungry) ate an average of 45% more food than the nonprimed children!

Patternicity is the overall tendency for humans to find meaningful patterns in meaningless stimuli, or to believe something is real when it is not. This often goes hand in hand with the concept of *agenticity*, which is the tendency humans have to see something as being "behind" the patterns we perceive (Shermer, 2011). These "invisible intentional agents" can help explain why children think nature is upset with them when their birthday party is rained out, or why adults are reluctant to put on clothing if told it belonged to a murderer. Philosopher David Hume (1757/1889) described this phenomenon in this way:

> There is an universal tendency among mankind to conceive all beings like themselves, and to transfer to every object, those qualities, with which they are familiarly acquainted, and of which they are intimately conscious. We find human faces in the moon, armies in the clouds; and by a natural propensity, if not corrected by experience and reflection, ascribe malice or good-will to every thing, that hurts or pleases us.

When we put priming and patternicity together with agenticity and our cognitive biases, one can start understanding why we not only see patterns around us, but why we see the particular patterns we do. When concerned parents were purposefully listening for messages in records being played backward, they thought that the musicians were trying to influence their children to worship the devil, have sex, and do drugs. So they were primed to hear certain words ("Satan," "sex," "marijuana"), which their naturally patternicity-prone brains were able to pick out from ambiguous, meaningless sounds. These "messages" were then assumed to have an agent behind them, purposefully attempting to subliminally manipulate impressionable youth. Finally, "findings" played into a preexisting confirmation bias and so were accepted quickly and unquestioningly ("See Harold, I told you Led Zeppelin was a bad influence on little Suzy!"). So, we see what it is we expect we are going to see, relying on what we've seen in the past and our brain's tendency to make patterns out of nothing. Using this framework, it's easy to explain why the unidentified flying object (UFO) enthusiasts want the Face on Mars to be real, or why the religious see their particular gods or goddesses on the side of overpasses or in the coat of a sheep, or how a sports fan could be sucked into believing in magical curses. But, given that this is a book on critical thinking, what does the empirical evidence, not just our outlined theory, say about these particular examples of pareidolia? How can we use the critical-thinking skills we learned in the previous chapter to examine these claims?

Backmasking Unmasked

To examine the fears that many people expressed about backmasked messages in popular music, we have to answer two separate questions: (a) were the musicians purposefully inserting such messages? and (b) could such messages have any impact on a person's behavior? As for the first one, the answer appears to be a resounding "no." All of the artists accused of inserting the hidden, backward lyrics into their songs resoundingly denied it . . . but of course those devilish demagogues would deny doing such a deed if they were doing it on purpose. We do have plenty of examples of musicians purposefully inserting backmasked messages or samples into songs since the 1960s, with the Beatles, Pink Floyd, Iron Maiden, the Electric Light Orchestra, Slipknot, Daft Punk, and even Weird Al[3] getting in on the act. In more recent years, Brittany Spears and Eminem have been accused of inserting backmasked messages into their songs. But, even if done on purpose and for nefarious purposes, could such a thing influence our behavior?

Beginning in the 1980s, researchers began trying to determine whether these messages could be understood, either consciously or subconsciously, and wheher they would impact behavior. In a classic study, Vokey and Read (1985) found that people were generally able to identify the gender of a speaker when listening to a voice played backward, but that was about it. People were unable to "decode" speech that was purposefully played backward, but once primed to hear certain phrases (such as "Saw a girl with a weasel in her mouth" if the Lewis Carroll poem "Jabberwocky" was played backward) people hear them very reliably. Follow-up research confirmed these findings, repeatedly showing no influence from such messages on overt behavior and a lack of ability to even detect the messages without being primed for them.[4] So, are bands putting hidden messages in your music, making you do things against your will? The answer is a resounding no.

Jesus, Martian, and Joseph!

So what about all those sightings of Jesus and Mary on tortilla chips, toast, clouds, office buildings, pizza pans, and the like, or the Face on Mars? This is a great example of the convergence of cognitive biases, priming, patternicity, and agenticity. Let's look back at the example of the Mary in the underpass discussed previously. Thousands of people passed by the stain on the wall daily, ignoring it completely, until Delgado, a devout Catholic, happened to be driving by it while praying. She was, therefore, massively primed to see not just a person's shape, but the shape of a particular religious icon. When this is added to our natural inclination toward patternicity (especially for human faces and forms), it helps explain why she was so moved by what she thought she saw. Then, by telling others about her interpretation of the water stain, she primed numerous others to "see" the same thing. Agenticity is also clearly at play, as Delgado (and many others) then assumed that such a thing had to have a force behind it, with many describing it as "miraculous" or a sign from God.

---

[3] A personal favorite is when on the song "Nature Trail to Hell," you can hear Weird Al say the phrase "Satan eats Cheese Whiz" if you play the song backward. He also inserted the backmasked phrase, "Wow, you must really have a lot of time on your hands" in the song "I Remember Larry."

[4] Confusingly, the people who claim there is an impact often point to a completely different area of psychological research to support their claim—that of subliminal perception. This is similar to the priming effects discussed earlier in this chapter, in which being briefly exposed to a particular stimuli can cause a change in subsequent behavior. In a review of the literature on the impact of subliminal persuasion techniques, Randloph-Seng and Mather (2009) conclude that such things only work in the context of being already motivated to do what the subliminal message is encouraging and a complete lack of awareness that the message is present.

Such people were then very reluctant to accept the explanation of pareidolia as a reason why it was there, instead falling prey to their confirmation biases.

The Martian face is another great example of this. When the original, low-resolution photo from 1976 was released to the public, NASA's press release on July 31 included a caption that stated:

> The picture shows eroded mesa-like landforms. The huge rock formation in the center, which resembles a human head, is formed by shadows giving the illusion of eyes, nose and mouth. The feature is 1.5 kilometers (one mile) across, with the sun angle at approximately 20 degrees. The speckled appearance of the image is due to bit errors, emphasized by enlargement of the photo.

The authors of the caption thought it would be a fun way to drum up excitement about the Viking 1 project, engaging the public while at the same time explaining why it looked like a face (i.e., the shadows at a particular angle and image errors). Instead, as we learned earlier, many people jumped on the Face as being evidence of intelligent life on Mars, and over the next 25 years it became the subject of an enormous amount of attention. Again, we saw that people became easily primed to see a pattern where there actually wasn't one (a face carved into a mesa), they then assumed some force or agent behind the pattern, and then we are off to the pseudoscientific theory race. Interestingly, in this case, the pareidolia actually spread to the surrounding Martian landscape, with people reporting that they could see details of numerous buildings, pyramids, and more objects in the Cydonia area.

With the unveiling of very high-resolution photos from the Mars Global Surveyor mission in the late 1990s and early 2000s that very clearly show a natural rock formation with no evidence of a face carved into it (NASA, 2001), we saw confirmation biases swing into full gear. Proponents of the Face ignored the new, sharper images (as seen in Figure 5.3) and instead focused on the older ones that were more ambiguous, dismissing the new photos and even claiming the originals showed evidence of large drawings (similar to the Nazca Lines in Peru) of primates and canines! Little in the way of hard evidence, it seems, can dissuade the true believer once a belief is embedded.

Statistical Unjinxing

For the Madden Curse and *Sports Illustrated* (*SI*) jinx, we can rely on statistical calculations to help in determining if these are true or just pareidolia. Sports statistician Rudy (2013) decided to apply some simple analyses to see whether there was a significant decline in performance for all the players who were on the Madden NFL covers between the years of 1999 and 2012. He was able to determine that, on average, the athletes played about

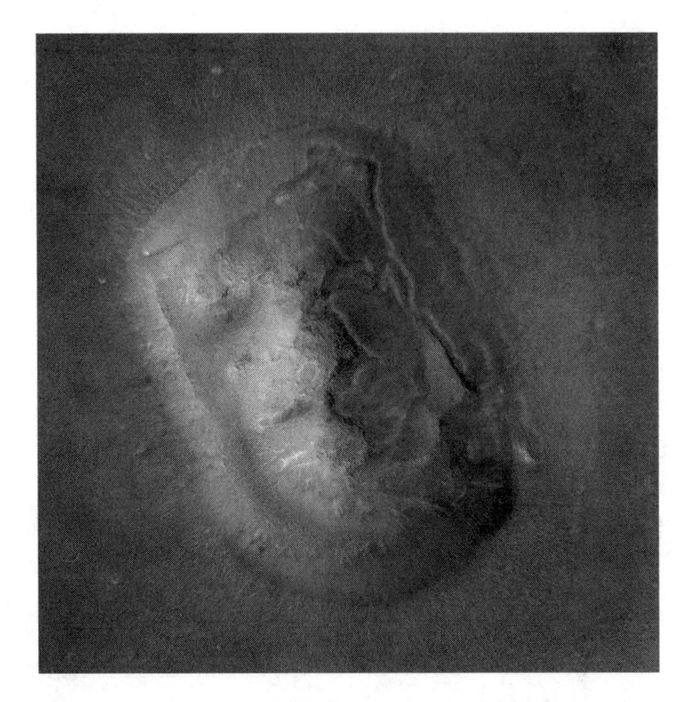

FIGURE 5.3 High resolution image of "Face on Mars".
Source: Courtesy of National Aeronautics and Space Administration (NASA). Image captured by the Mars Global Surveyor on April 8, 2001.

two fewer games the season after appearing on the cover, but that this average was massively lowered by a single player breaking his leg and missing almost the entire season. Performance-wise, he found that there was a significant decline in performance the season after someone was on the cover, but that there was no difference in their postcover performance and their average performance the two seasons before the "cover season." In other words, the season that got them on the cover was the unusual one, the outlier.

The Madden Curse is a classic case of what we call *regression to the mean*. Think of flipping a coin; overall, there is a 50% chance that each flip will be heads. But sometimes you might get five heads in a row, or five tails. Flip it long enough, though, and you'll have an overall average of 50% heads. In other words, even if you have an unusual run of heads that appears out of the ordinary, the overall pattern will be the same in the long run. For our football players, the reason they were selected for the cover in the first place is that they had an extremely good performance in a particular season. When their performance the next season was not at the same high level, people them as being "cursed," but in reality their performance just went back to normal levels, or regressed to the mean.

The *SI* jinx is also easily examinable statistically. We would simply need to determine how many covers have been published since the magazine's launch and how many "jinxed" athletes there have been to see what the odds are that a player would have bad luck after being on the cover. The most extensive list that your authors could find indicated 115 "notable incidents" of the *SI* jinx since 1954. At first, this sounds like a very large number of times for this to happen . . . until you consider how many covers there have been in the 60 years of publication. *Sports Illustrated* has been published as a (mostly) weekly magazine since 1954, and a total of 3,069 covers came out during the first 60 years of publication. This means that out of 3,069 covers, only 3.8% (115) have "caused" bad luck. Not exactly a large percentage. In fact, the person with the record for most cover appearances, basketball superstar Michael Jordan, had no incidents of the "jinx" in his 50 times on the cover. This is a great example of post-hoc (after the fact) reasoning (we see someone who has been injured, had a bad game or season, and then seek out reasons why that happened) and confirmation bias at play (we ignore all the times someone was on the cover and performed well, but remember the ones that confirm our belief in the jinx).

CONCLUSIONS

We hope by now you are starting to get the idea that we can be easily fooled by our brains into believing things that clearly go against the evidence at hand. Although there is no way to completely rid ourselves of these cognitive biases, there are a number of tools and methods we can use to mitigate their effects on our everyday decision making. First, doing what we have just done (learning about the biases) is called awareness raising. It is certainly useful in assisting us in recognizing these biases when they manifest in others' decisions, but unfortunately does little to prevent us from falling prey to them.

This lack of personal insight appears to be primarily attributable to the fact that these are largely automatic processes that manifest across almost all aspects of our cognition (Gilovich, 1991; Schacter, 1999). Indeed, some evolutionary theorists posit that these are not actually cognitive flaws because many of these processes save time and effort and do not always produce poor decisions and judgments. It is argued that these biases and information-processing problems were selected for during our evolutionary history and may be better regarded as resource conserving, utilitarian, mental shortcuts (Haselton et al., 2009). Similarly, patternicity appears to be a highly useful tool, evolutionarily speaking. Being able to discern patterns in the movement of animals can make you a better hunter. Quickly recognizing potential threats in your environment, like snakes or spiders, allows you to react more easily to them and avoid harm.[5] Likewise, attributing agency to random events may be a useful way to feel more in control of one's surroundings by thinking that you can influence those events (e.g., by making appropriate sacrifices to the volcano spirit, it won't erupt and destroy us all). But as we've seen in the many preceding examples, these natural tendencies don't always lead us to accurate beliefs about the world around us.

Nonetheless, having awareness can at least lead us to engage in some cognitive self-reflection on our decisions and automatic impulses, causing us to learn to doubt even our own minds and avoid some of the biases and heuristics (Toplak, West, & Stanovich, 2010) described. Many people fail to doubt, instead falling into what is called the bias blindspot, recognizing biases in others but reporting that they are less prone to them. Self-reflection alone is not enough though, as research has also shown that those who are free of the bias blindspot are not automatically unshackled from their own cognitive biases (West, Meserve, & Stanovich, 2012).

If awareness and reflection are not enough, what else can we do? Psychologists have recognized specific methods of decision making that may help to mitigate the effects of our mental shortcuts. The most prominent include training in critical-thinking skills, which often emphasize honing one's ability to evaluate evidence independently of one's previously held opinions or beliefs (Sternberg, 2001). Relying on empirical evidence, rather than intuition and "snap" judgments, is another way to avoid potential biases. Allowing empiricism to guide beliefs and decisions to the best degree possible is useful because the methods of science (as discussed in Chapter 2) are specifically designed to minimize potential biases in hypothesis testing (unlike the human brain).

In conclusion, we (and everyone we encounter in life) will always have cognitive biases and heuristics as part of our mental toolbox. They will often be useful, but can also lead us to make systematic mistakes in decision making and judgment. Therefore, we can and should try to battle them. Learning to doubt our brains is step number one in this process, but it is only the first step in becoming a more effective and less biased decision maker. Training ourselves to think critically, while also employing the use of solid problem-solving steps like the scientific method, can then allow us to be less prone to the biases and heuristics reviewed previously.

---

[5] In an increasingly well-supported hypothesis, anthropologist Lynne Isbell puts forth the idea that the highly developed visual acuity displayed by primates, including humans, is the result of an "evolutionary arms race" with snakes—our ancestors developed better vision and even specialized areas of the brain to detect snakes and avoid being eaten (Isbell, 2011).

QUESTIONS FOR REFLECTION

1. *To what extent do you engage in metacognition, or thinking about your own thinking? Do you think that you should do more of this? How might you accomplish that?*
2. *What are the possible advantages of heuristics? Can you think of the last time you applied a heuristic? (And are you perhaps doing so right now, in response to this question?)*
3. *What does belief perseverance tell us about how we should respond to well-intentioned criticism offered by friends and family?*
4. *Imagine that a film producer has had a string of successful movies, each breaking box-office records. If his next two movies flop at the box office, is that good reason for the studio to terminate his contract?*

REFERENCES

Anderson, C. A. (2007). Belief perseverance. In R. F. Baumeister & K. D. Vohs (Eds.), *Encyclopedia of social psychology* (pp. 109–110). Thousand Oaks, CA: Sage.

Ariely, D. (2009). *Predictably irrational: The hidden forces that shape our decisions* (1st ed.). New York, NY: HarperCollins.

Azevedo, F. A. C., Carvalho, L. R. B., Grinberg, L. T., Farfel, J. M., Ferretti, R. E. L., Leite, R. E. P., . . . Herculano-Houzel, S. (2009). Equal numbers of neuronal and nonneuronal cells make the human brain an isometrically scaled-up primate brain. *Journal of Comparative Neurology, 513*(5), 532–541.

Benson, H., Dusek, J. A., Sherwood, J. B., Lam, P., Bethea, C. F., Carpenter, W., . . . Hibberd, P. L. (2006). Study of the therapeutic effects of intercessory prayer (STEP) in cardiac bypass patients: A multicenter randomized trial of uncertainty and certainty of receiving intercessory prayer. *American Heart Journal, 151*(4), 934–942.

Blank, H., Nestler, S., von Collani, G., & Fischer, V. (2008). How many hindsight biases are there? *Cognition, 106*, 1408–1440.

Combs, B., & Slovic, P. (1979). Newspaper coverage of causes of death. *Journalism Quarterly, 56*, 837–849.

Eder, A. B., Fielder, K., & Hamm-Eder, S. (2011). Illusory correlations revisited: The role of pseudocontingencies and working-memory capacity. *Quarterly Journal of Experimental Psychology, 64*(3), 517–532.

Epley, N., & Gilovich, T. (2006). The anchoring-and-adjustment heuristic: Why the adjustments are insufficient. *Psychological Science, 17*, 311–318.

Gigerenzer, G., & Gaissmaier, W. (2011). Heuristic decision making. *Annual Review of Psychology, 62*, 451–482.

Gilovich, T. (1991). *How we know what isn't so: The fallibility of human reason in everyday life.* New York, NY: The Free Press.

Harris, J. L., Bargh, J. A., & Brownell, K. D. (2009). Priming effects of television food advertising on eating behavior. *Health Psychology, 28*(4), 404–413.

Haselton, M. G., Bryant, G. A., Wilke, A., Frederick, D. A., Galperin, A., Frankenhuis, W. E., & Moore, T. (2009). Adaptive rationality: An evolutionary perspective on cognitive bias. *Social Cognition*, *27*(5), 733–763.

Hume, D. (1889). *Natural history of religion*. Online Library of Liberty, Liberty Fund, Inc. (Original work published 1757). Retrieved from http://oll.libertyfund.org/titles/340

Isbell, L. A. (2011). *The fruit, the tree, and the serpent: Why we see so well*. Cambridge, MA: Harvard University Press.

Kahneman, D., Slovic, P., & Tversky, A. (1982). *Judgment under uncertainty: Heuristics and biases* (1st ed.). Cambridge, UK: Cambridge University Press.

Kahneman, D., & Tversky, A. (1972). Subjective probability: A judgment of representativeness. *Cognitive Psychology*, *3*(3), 430–454.

Lebovich, J. (2005, April 19). Faithful see Mary on underpass wall. *Chicago Tribune*. Retrieved from http://www.chicagotribune.com/chi-0504190294apr19-story.html

Marshall, R. (2014). *Confidence or curse? Looking back on Madden's troubled cover athletes*. Retrieved from http://www.digitaltrends.com/gaming/the-madden-curse

Marsolek, C. J. (2003). What is priming and why. In J. S. Bowers & C. J. Marsolek (Eds.), *Rethinking implicit memory* (pp. 41–64). Oxford, UK: Oxford University Press.

Masters, K. S., Spielmans, G. I., & Goodson, J. T. (2006). Are there demonstrable effects of distant intercessory prayer? A meta-analytic review. *Annals of Behavioral Medicine*, *32*(1), 21–26.

National Aeronautics and Space Administration. (1976, July 31). Viking News Center, press release. Pasadena, CA: NASA. Retrieved from http://www.msss.com/education/facepage/pio.html

National Aeronautics and Space Administration. (2001). *Unmasking the face on Mars*. Retrieved from http://science.nasa.gov/science-news/science-at-nasa/2001/ast24may_1

Nickerson, R. S. (1998). Confirmation bias; A ubiquitous phenomenon in many guises. *Review of General Psychology*, *2*(2), 175–220.

Plous, S. (1993). *The psychology of judgment and decision making*. Columbus, OH: McGraw-Hill.

Randolph-Seng, B., & Mather, R. D. (2009). Does subliminal persuasion work? It depends on your motivation and awareness. *Skeptical Inquirer*, *33*(5), 49–53.

Roese, N. J., & Vohs, K. D. (2012). Hindsight bias. *Perspectives on Psychological Science*, *7*, 411–426.

Rudy, K. (2013). *Is the "Madden curse" real?* Retrieved from http://blog.minitab.com/blog/the-statistics-game/is-the-madden-curse-real

Sacks, O. (2012). *Hallucinations*. New York, NY: Vintage.

Schacter, D. L. (1999). The seven sins of memory: Insights from psychology and cognitive neuroscience. *American Psychologist*, *54*(3), 182–203.

Schwarz, N., Bless, H., Strack, F., Klumpp, G., Rittenauer-Schatka, H., & Simons, A. (1991). Ease of retrieval as information: Another look at the availability heuristic. *Journal of Personality and Social Psychology*, *61*, 195–202.

Shermer, M. (2011). *The believing brain: From ghosts to gods to politics and conspiracies—How we construct beliefs and reinforce them as truths*. New York, NY: St. Martin's Griffin.

Sternberg, R. J. (2001). Why schools should teach for wisdom: The balance theory of wisdom in educational settings. *Educational Psychologist, 36*, 227–245.

Strack, F., & Mussweiler, T. (1997). Explaining the enigmatic anchoring effect: Mechanisms of selective accessibility. *Journal of Personality and Social Psychology, 73*(3), 437–446.

Toplak, M. E., West, R. F., & Stanovich, K. E. (2011). The cognitive reflection test as a predictor of performance on heuristics and biases tasks. *Memory and Cognition, 39*, 1275–1289.

Vokey, J. R., & Read, J. D. (1985). Subliminal messages: Between the devil and the media. *American Psychologist, 40*(11), 1231.

West, R. F., Meserve, R. J., & Stanovich, K. E. (2012). Cognitive sophistication does not attenuate the bias blind spot. *Journal of Personality and Social Psychology, 103*(3), 506–519.

Zusne, L., & Jones, W. (1989). *Anomalistic psychology: A study of magical thinking* (2nd ed.). Hilsdale, NJ: Lawrence Erlbaum Associates.

CHAPTER 6

WHY CAN'T WE TRUST OUR WORLD?

It's easy to lose objectivity when we feel strongly about an issue. Some of the things we feel strongly about might also be of great consequence—for example, our personal health or the health of the planet—making it even more difficult to separate the strength of your emotional commitment from the strength of your argument. Unfortunately though, the more important something is, the more important it *also* becomes that our reasoning be sound, so that we can stand a better chance of convincing others that our viewpoint is well-justified. Being as rational as we'd like to be is never easy, though, especially when we live in an age of celebrity-as-automatic-authority, short attention spans, and a lack of patience for complex arguments.

What this adds up to is elegantly illustrated in one of Randall Munroe's (n.d.) excellent xkcd.com comics, reproduced as Figure 6.1. The point is that unless a claim is made fairly hyperbolically (what's often referred to as "clickbait" online), the chances are good that it will get far less attention than it might otherwise merit.

Once hyperbole becomes the norm (and perhaps that particular horse has already left the stable), the space for careful deliberation narrows, and the quality of our conclusions will likely decline. This is one way in which the external world—or rather, information we're given about the external world, especially in popular media—can contribute to shoddy thinking. This chapter explores these and other ways in which our desire to think carefully and critically can be thwarted, or at least complicated, by factors mostly beyond our control. Fortunately, though, being aware of these factors can be invaluable in minimizing their negative impact on our reasoning.

CONFIRMATION BIAS AND MOTIVATED REASONING

Confirmation bias, as discussed extensively in the last chapter, is our predisposition to grant disproportionate positive value to evidence that supports our existing beliefs, while dismissing or undervaluing contrary evidence (Nickerson, 1998). Whenever we make inductive inferences (see Chapter 2), we extrapolate from available evidence to a conclusion—but with confirmation bias, even if the available evidence is true, we interpret it and weigh it in

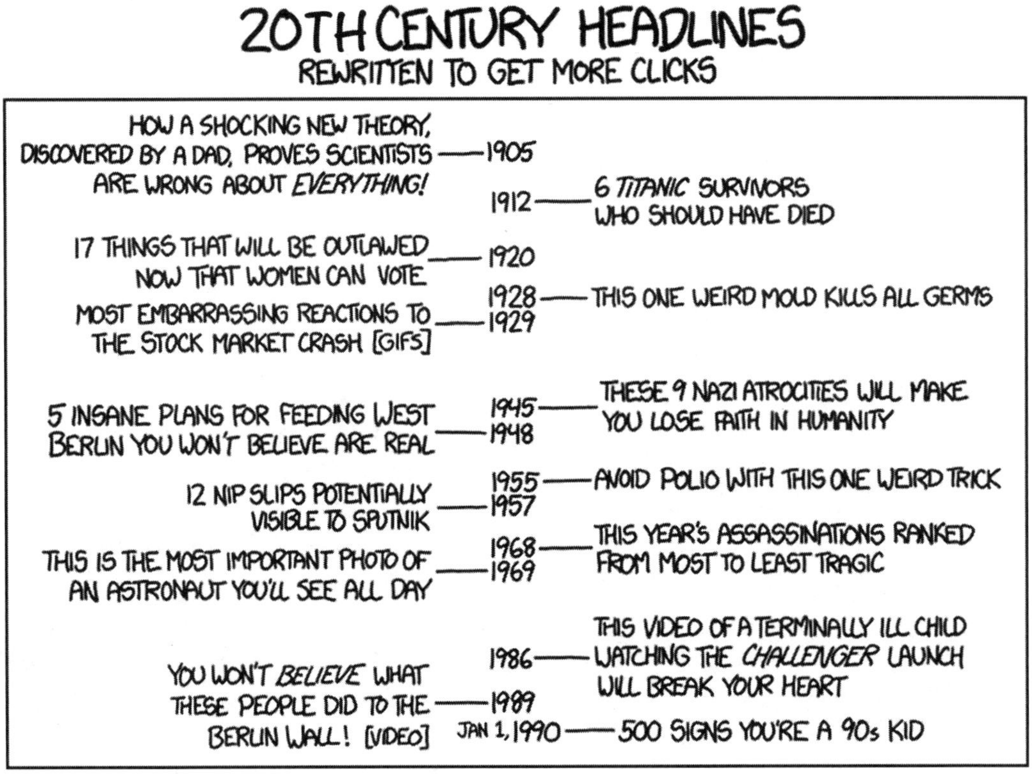

FIGURE 6.1 20th Century headlines

Source: Courtesy of xkcd, a Webcomic of Romance, Sarcasm, Math, and Language (https://xkcd.com/1283) under a Creative Commons Attribution-NonCommercial 2.5 License.

a manner that biases us to the conclusion we'd prefer to be true, rather than the one suggested by a fair assessment of the evidence.

Imagine that you were watching a political debate. If your preferred candidate stumbles over her argument, confirmation bias might lead you to think that she was asked a trick question, rather than that she was simply unprepared or didn't have a satisfactory answer. If you're an investor, your perception of your success in the market might be skewed toward remembering your successes (and thinking that you're on the road to riches), while not paying sufficient attention to your losses, which actually indicate that you might instead want to hand your portfolio over to a professional! And you will see in Chapter 9, if you visit a psychic you might be far more impressed with his "hits" (the accurate things he says about you) rather than his "misses" (inaccurate statements about you)—to the extent that you might not even allow yourself to notice that the "hits" were the sorts of things that anyone might have been able to guess without needing to communicate with a spirit world!

We're all prone to confirmation bias to varying degrees. After all, who wants to be wrong? Unfortunately, standing a better chance of being right *requires* that we embrace the fact that wanting to be right about what we believe can predispose us to error, and that we should instead do the best we can to remain alert to how we can fool ourselves, or become fooled through selective attention to that which is agreeable to us.

The reason this discussion is situated in a chapter that discusses why we can't trust our world is that even though we come to believe various things through personal experience and reflection, the vast majority of what we end up believing comes to us from external sources—what we read in the newspapers or on the Internet, what we hear on the news, or what people tell us. And once we know that they, like us, are prone to confounding factors like confirmation bias, this can serve as a reminder that it doesn't take purposeful ill intent for others in the world to end up deceiving us. They might inadvertently do so, simply because they have ended up believing something false via a nonobjective weighing of the available evidence.

To compound the problem, the lack of objectivity can go far deeper than simply our assessment of evidence. *Motivated reasoning* describes a substructure or foundation underlying our confirmation bias, with which we develop background rationalizations to justify holding our beliefs, even in the face of strong contrary evidence and argument. Conspiracy theorists often provide good examples of this, in that someone who holds a contrarian view such as theirs has a built-in argument as to why the evidence in support of the conspiracy is so weak—others are suppressing it, so it hardly counts against their view that it can't be satisfactorily demonstrated to be true—you're simply being unreasonably demanding!

THE BACKFIRE EFFECT

Unfortunately, there's even more bad news regarding how much faith we can have in people changing their minds in light of the best available evidence and arguments. As McRaney (2011) points out on the You Are Not So Smart website, "When your deepest convictions are challenged by contradictory evidence, your beliefs get stronger." McRaney is summarizing "the backfire effect," a phrase coined by Nyhan and Reifler (2010) in their research indicating that debunking misinformation is far more difficult than we might think (and hope) becaus when we present evidence that contradicts what someone else believes, that person is more likely to dig in and resist correction than to accept that he or she was wrong. So instead of facts serving to undermine false beliefs and misconceptions, they can serve to reinforce those errors. This is, of course, very bad news in a number of fields—climate change, politics, and health science are three that spring immediately to mind.

As Reifler suggests in an interview with Silverman (2011), we can find ways around the backfire effect, but we shouldn't imagine that refuting nonsense is easy, no matter how nonsensical or dangerous that particular piece of nonsense might be. Reifler points out that one might be able to get motivated reasoners to a tipping point where they are willing to change their minds, but perhaps only through persistent repetition of the correction you're trying to make. One or two headline news corrections might not do it, but weeks of drip-feeding the correction across all sorts of media might. Another possible strategy is suggested by Philip Fernbach of the University of Colorado, who published the results of an interesting 2013 experiment (Fernbach, Rogers, Fox, & Sloman, 2013), which that suggest that instead of providing reasons for why we think others are mistaken, we should instead try to provide *explanations.*

For example, instead of asserting that we need universal health care because everyone is morally equal in this respect, and therefore equally entitled to care from the state, try explaining how your envisaged universal health care scheme would work—how would it be implemented, what would it cost, who would pay, who would benefit? According to Fernbach, this approach stands a better chance of persuading others that you are right, because you have "shown your workings," rather than simply asserted your view, and in doing so have offered your opponent some of the resources that will help her to discover her errors through thinking the problem through, rather than just being told she is wrong. If you are simply offering reasons (rather than explanations), it's more likely that you'll appear interested in demonstrating that you're right rather than having a conversation or debate—and this might well cause your opponent to rebel against what you're saying. By contrast, if you're explaining, the research suggests that you would be better able to effectively articulate to others why they should be persuaded to subscribe to your point of view.

FILTER BUBBLES AND ECHO CHAMBERS

The final background issue to touch on before looking at an example of all these problems in operation is the *filter bubble* or *echo chamber*. The concept was initially popularized by Pariser (2011) in his TED (Technology, Entertainment, Design) talk titled "Beware Online 'Filter Bubbles,'" in which he highlighted how personalized news services (as well as ubiquitous tools such as Google search) can, over time, tell you more and more of what you *want* to hear rather than providing you with a neutral point of view. Geolocation data, as well as other personal data about things like your browsing habits, are stored in your web browser's cookies (small text files that record limited personal information about your browsing habits) and can influence the output of your web searches and other Internet browsing.

On first glance, this might not seem problematic—after all, don't we *want* to be exposed to themes that we are already familiar with, and know are of interest to us rather than not? Answering "yes" to this question might seem intuitively reasonable, but the problem is that your perception of what the consensus view is might quickly become skewed through only being exposed to like-minded communities and ideas. Consider the clear political and ideological divides between Fox News and CNN in the United States, or the *Daily Mail* and *The Guardian* in Britain. It's true that depending on whether you are liberal or conservative, you might *prefer* to consume news from one or another of these sources—but it's also true that you'll stand a greater chance of reading a caricatured or incomplete account of the arguments of those you might disagree with, and therefore stand a lesser chance of being persuaded by those arguments. If we all care to know the truth rather than that others share our positions, it's clear how this could present a problem.

The solution to avoiding the worst possible influences of the filter bubble starts with knowing that you might find yourself inside one, thanks to the sources you allow yourself to be exposed to. If you are aware of that possibility you can make an effort to expose yourself to different points of view, whether on the Internet or in conversation. This is perhaps especially important on social media, as the number and volume of voices you can encounter on places like Twitter and Facebook can encourage broadcasting rather than reflective comment, with opposing voices easily dismissed as not only wrong, but also sometimes abusively so in the case of Internet "trolls" (Pullen, 2015). The simplest and most effective advice, though, for avoiding filter bubbles might come from Alan Martin (2013), who says: "Perhaps the best way though is to directly counter your instincts: click on links you may not like, and read the comments." As painful as his final suggestion might be, this attempt to find disconfirming evidence for your own views is much more in line with the skeptical and scientific viewpoints of a critical thinker.

To amplify the importance of clicking on links you may not like—rather than the ones from sources you might already know and trust—consider also Fisher, Goddu, and Keil's (2015) finding that "searching the Internet for explanatory knowledge creates an illusion whereby people mistake access to information for their own personal understanding of the information" (p. 674). In other words, searching for and reading information related to a debate might create the impression that you understand its complexities, even though all you might have discovered is simply that the information exists. This awareness of the existence of an answer is, of course, a far less useful thing than awareness of what the actual answer is—even more so if the former can operate as an illegitimate substitute for the latter.

MORAL PANICS AND DIETARY HYPERBOLE

Our health is a topic that can lend itself to overreaction. In particular, the sort of overreaction that treats debates as resolved when that is the furthest thing from the truth. But the easy answer or the quick cure—often to be found in "fad diets"—tells a compelling story, gives us a ready-made answer to our problems, and also frequently offers us a scapegoat to blame for our woes. Sometimes these dietary movements seem similar to religious cults, with insider knowledge ("this secret ingredient is killing you!"), particular rituals ("here's a great cooking tip to halve the calories in your rice!"), and sometimes even prophets who can lead us from the wilderness of confusion, so long as you trust and obey. The fact that we all *do* tend to regard our health as a good in and of itself, and a very important one at that, can lead to our being susceptible to discarding nuanced—and more accurate—understandings of the scientific process and its conclusions in favor of misleading headlines and hyperbole.

One of the important lessons health care professionals, scientists, and science writers can teach others, including the public and government policy makers, is that things are often uncertain. But uncertainty doesn't sell books or newspapers, and is also a far less compelling conversation to have over dinner than one about the latest miracle cure. Furthermore, when debates on scientific matters express certainty, or when scientists adopt a dogmatic stance, it can not only foreclose debate but more important, can put science in the same realm as pseudoscience. Homeopaths confuse the public with unfalsifiable claims, astrologers do likewise, and these things waste people's time, money, and, occasionally, lives. It's the fact that critical thinkers embrace questions and uncertainty that makes the scientific method superior as a means of evaluating information, because it keeps us alert to our potential errors.

To put it another way, being right often starts with embracing the possibility that you might be wrong. By contrast, the tone of much popular discourse, including coverage of important scientific fields in newspapers and on social media, proceeds as if things can be known for certain. This leads to absurd contestations with things being "proved" and then "disproved" with each new bestseller, and apparent "authorities" rise and are then quickly forgotten as our attention shifts to the next sensation. This infantilizes the public—not only by treating them as if they are unable to make choices for themselves, but also more literally, in helping to *ensure* that they can't make choices for themselves by misleading them and teaching them to believe in simplified versions of the truth.

What can be forgotten here is that *thinking* is a good in itself. Diet and health care choices are just one area in which we need to make decisions, and rational decisions are only possible if we are thinking things through, in full awareness of how the easy answers might mislead us. For a contemporary example of how hyperbole can triumph over careful (and more accurate) thinking, let's consider the current moral panic around sugar. Sugar, according to some (Hyman, 2014), is a substance that is "more powerfully addictive than alcohol, cocaine or even heroin." This message has found a receptive audience in the cinema, with two 2014 films (*Fed Up* and *That Sugar Film*) warning about the food industry's role in making us fat and sick, and in particular, sugar's insidious contribution to the problem.

Is Sugar Addictive?

The short and simple answer is that at this point in time, we don't know. Yet, depending on how you choose to interpret a word like "addiction," you might be able to construct a plausible case for either a "yes" or a "no" answer to that question. There are various reasons to be suspicious of any glib answer to this question, partly because of a point made by Swiss-German physician Paracelsus in the 15th century: "The dose makes the poison." In other words, the fact that some people might have an unhealthy relationship with sugar might stem from the fact that they consume far too much of it, and this is not at all the same thing as saying that the substance in itself, is a toxic or addictive one. Although there may be no safe number of cigarettes to smoke, there will be a dosage of sugar that's unproblematic in all but the rarest of cases. Is it then accurate to say that sugar is addictive, even if most of us can eat a limited quantity of it without suffering any harm?

When the average person thinks about something as being addictive, she means that the substance or activity in question is *particularly* likely to cause you to develop some combination of dependence, tolerance, cravings, and withdrawal symptoms. You'd also, if addicted, go to significant lengths to obtain the thing that you are addicted to. On a trivial end of this spectrum, people who smoke cigarettes might walk out into a cold and rainy evening to go and purchase cigarettes instead of staying under the blanket on the couch like any sensible (i.e., nonaddicted) person would.

When you think of addiction, in other words, you typically don't mean that you know this fellow, George, who has become so obsessed with playing Minecraft that you describe Minecraft as addictive. Instead, you might acknowledge that people can become "addicted" to Minecraft or, in other words, particularly obsessed with it because of *their* dispositions, rather than because the thing in question is particularly *conducive* to dependence.

The distinction is important, and points to one of the significant problems in discourse around sugar "addiction." It's important because the things we like are rewarding at the level of the brain, in that they result in dopamine release to the nucleus accumbens (Harvard HelpGuide, 2015). A faster and stronger dopamine release suggests greater pleasure, and once one develops a tolerance for the "reward" in question (which could be cocaine, sugar, sex, running, or whatever else you enjoy) can also indicate addiction. But when we talk about *things* being addictive, we use the term in its stronger sense, outlined previously, meaning that the average person *can't help* but find that thing rewarding, or will typically tend to start to find it rewarding, even if he started out disinterested. You'll be able to imagine your own examples based on your particular circumstances, but in the case of your authors, there's almost zero chance of our becoming addicted to running because we don't like doing it and tend not to do it. But there's a significant chance—and a similar chance to yours chance—of our becoming addicted to heroin if either of us were to try it.

We're using the word "addictive" in a very broad sense when we describe the Internet, exercise, and sugar as addictive. In fact, the sense in which it's being used is broad enough to lose significant meaning. Most of us exercise, ingest sugar, use the Internet, and have sex quite unproblematically. Interestingly, more of us use things like heroin or cocaine without the dire consequences that the standard sort of panic around addiction would suggest, despite the human tragedies that can often accompany these substances.

One compelling example of this recounted by Hari (2015) is that of President Nixon's "Operation Golden Flow." In 1971, there was an epidemic of heroin use in Southeast Asia among U.S. soldiers. To attempt to combat this, Nixon launched Operation Golden Flow, which entailed soldiers being barred from boarding a plane back to the United States unless they tested clear of heroin. Given our perception of heroin as one of the more addictive drugs, and one of the most difficult to give up, the results were remarkable. In July of that year, an estimated 15% of soldiers were using heroin regularly, but 2 months later, all but 4.5% tested clear. In follow-up studies, some 3 years later, only 12% of those who had previously displayed heroin dependence had reexperienced addiction.

These and other data show us that long-term addiction is the exception, not the rule—we suffer from a confirmation bias here in the sense that we don't get to hear about the people who live with addictions that are largely under control. We hear about the horror stories of people struggling with a demon and (sometimes) heroically fighting it off, and these accounts can interfere with a more balanced understanding of how people consume addictive substances. So one can agree that a problem like "sugar addiction" is overstated while also agreeing that people are eating too much sugar, and that it's a problem when they do so unknowingly. For example, eating foods with undisclosed added sugars, which can fool us into thinking we are consuming a healthier diet than we actually are, is a problem.

It's also true that people who are prone to compulsive behavior might well find themselves becoming addicted to sugar—but for some (and perhaps, all) of those people, the problem might be with their lives and their circumstances (at least in large part) rather than with the substance itself. We should not be surprised that our brains find food rewarding and that we seek it out. We would be surprised if it was any other way. But if we can (typically, as with most consumers of sugar) control the impulse to eat too much of it, then addicts need to shoulder a large portion of the responsibility themselves, and not hand it over to sugar. So why is it that the less nuanced "sugar is addictive" narrative has gained such popularity?

PRESSURE TO PUBLISH SCIENTIFIC RESEACH AND PRESS RELEASES

Part of the problem is that sensationalism sells, as it always has. Another part of the problem is that it is fairly easy to find something to sensationalize, as John Bohannon demonstrated with a 2015 hoax in which he managed to fool many serious publications (*Bild*, *Cosmopolitan*, and the *Huffington Post*, to mention but three) and many hundreds of thousands of people into believing that chocolate can contribute to weight loss (Bohannon, 2015). The paper has since been retracted, with the journal in question claiming it was never really published. You can read about that element of the story on Retraction Watch (Oransky, 2015), but for our purposes, what Bohannon did is particularly interesting in that it wasn't a hoax in the classic sense—he did in fact find the results that his paper claimed to have found. Here's how he explains this:

> Here's a dirty little science secret: If you measure a large number of things about a small number of people, you are almost guaranteed to get a "statistically significant" result. Our study included 18 different measurements—weight, cholesterol, sodium, blood protein levels, sleep quality, well-being, etc.—from 15 people. (One subject was dropped.) That study design is a recipe for false positives.
>
> Think of the measurements as lottery tickets. Each one has a small chance of paying off in the form of a "significant" result that we can spin a story around and sell to the media. The more tickets you buy, the more likely you are to win. We didn't know exactly what would pan out—the headline could have been that chocolate improves sleep or lowers blood pressure—but we knew our chances of getting at least one "statistically significant" result were pretty good.

Now, his study did show accelerated weight loss in the group that ate a low-carbohydrate diet *and* a bar of chocolate versus a group that simply ate the low-carbohydrate diet, but only because of a small sample size (which tends to exaggerate trivial findings). And when those findings tell us something particularly interesting, attractive, or controversial, they can be published with very little oversight in terms of whether the results are credible or not, or more important, given how easy it can be to find a statistically significant result, whether the results are genuinely *illuminating*—telling us something that should change our behavior. To quote Bohannon again,

> Take a look at the press release I cooked up. It has everything. In reporter lingo: a sexy lede, a clear nut graf, some punchy quotes, and a kicker. And there's no need to even read the scientific paper because the key details are already boiled down. I took special care to keep it accurate. Rather than tricking journalists, the goal was to lure them with a completely typical press release about a research paper. (Of course, what's missing is the number of subjects and the minuscule weight differences between the groups.)

This problem isn't only felt with hoaxes such as this. The hoax instead highlights a widespread problem discussed by Sumner et al. (2014) in the *British Medical Journal*, who found that even though it's true that newspapers and other public media often contain exaggerated descriptions of research findings, the exaggerated advice and causal claims carried in the newspapers are frequently to be found in press releases themselves—even when those press releases are written by the press office of the university responsible for the research! Dr. Ben Goldacre's editorial (2014) highlights the extent of the problem:

> Sumner et al. (2014) identified all 462 press releases on health research from 20 leading UK universities over one year. They traced 668 associated news stories and the original academic papers that reported the scientific findings. Finally, they assessed the press releases and the news articles for exaggeration, defined as claims going beyond those in the peer reviewed paper.
>
> Since coding for exaggeration could be subjective, the authors' structured appraisal focused on three areas: making causal claims from correlational findings in observational data, making inference about humans from studies on other animals, and giving direct advice to readers about behaviour change. This allowed an assessment of where each exaggeration first appeared. If a news story claimed a new treatment for humans, for example, but the study was on mice—and the academic paper made no claim about humans—then did the exaggeration first appear in the press release or the newspaper article?
>
> Over a third of press releases contained exaggerated advice, causal claims, or inference to humans. When press releases contained exaggeration, 58% to 86% of derived news stories contained similar exaggeration, compared with exaggeration rates of 10% to 18% in news articles when the press releases were not exaggerated.

In other words, it's not just the media's fault. Under pressure to secure funding and bolster reputations, scientists themselves—and their press officers—can be guilty of exaggerating their findings in order to make them more attractive to a public that is hungry for sensation or the next miracle cure. Nevertheless, a portion of the blame can still sometimes be apportioned to journalists who put the most seductive spin on the research in question as well as on us, who might be primed to interpret things according to what we'd like them to say, rather than to accept that the reality is that not all scientific findings are equally compelling, and that we should be suspicious when they tend to be treated as if they are. A salutary example of how media representation and our willingness to believe can influence perception can be found in an oft-cited paper by Avena, Rada, and Hoebel (2008), who tell us that:

> Food is not ordinarily like a substance of abuse, but intermittent bingeing and deprivation changes that. Based on the observed behavioral and neurochemical similarities between the effects of intermittent sugar access and drugs of abuse, we suggest that sugar, as common as it is, nonetheless meets the criteria for a substance of abuse and may be "addictive" for some individuals when consumed in a "binge-like" manner. This conclusion is reinforced by the changes in limbic system neurochemistry that are similar for the drugs and for sugar. (p. 30)

Given that this is a paper that is often claimed to support the idea that sugar is, in fact, addictive, a critical reader might (or perhaps should) immediately be asking "who might be inclined to consume in a 'binge-like' fashion?" and then follow that up with the hypothetical answer that if a person has some preexisting impulse control disorder, or just poor self-discipline when it comes to candy, he might well latch on to sugar as an *expression*, rather than a *cause*, of that lack of impulse control.

Also, notice the scare quotes around "addiction," through which the authors seem to be hedging their bets, with the text itself only weakly supportive of any claim to sugar addiction, at least if "addiction" is taken to mean what it normally does (dependence, craving, withdrawal, and so forth). This reading is amplified by a later quote, which reads:

> It is not clear from this animal model if intermittent sugar access can result in neglect of social activities as required by the definition of dependency in the *DSM-IV-TR* (American Psychiatric Association, 2000). Nor is it known whether rats will continue to self-administer sugar despite physical obstacles, such as enduring pain to obtain sugar, as some rats do for cocaine. (p. 30)

To paraphrase, "there are some fairly typical features of what we normally understand as addiction that are missing here, yet there's the possibility that the word *addiction* might nevertheless be a good fit." Now, I don't mean to impugn Avena et al.'s research—I use the example of this paper simply to make the point that interpreting it as *proof* that sugar is "addictive" (Langreth & Stanford, 2011) requires you to adopt a certain definition of what "addictive" means in the first place, and that definition is a highly contested one. The word "addiction" might mean something quite different to the parent whose child has a heroin habit than it would to a scientist in a lab—but those differences are glossed over in popular media coverage of the research (and, unfortunately, in much of the research itself). With addiction, one is perhaps reminded of a line from Lewis Carroll's *Through the Looking Glass* (1991), where Humpty Dumpty said: "When I use a word, it means just what I choose it to mean—neither more nor less."

The hyperbole in blogs and online news sources, never mind repositories of the worst sorts of pseudoscience like *Natural News*, don't help to clarify these issues. Neither do personal anecdotes, regardless of our compassion for people who struggle with compulsive behavior of various sorts. Movies like *Fed Up* are of little use also, in that they simply populate a panicked filter bubble with cherry-picked and misrepresented data (Hall, 2014).

We don't yet have good human data for sugar addiction. What we do seem to have is limited evidence for "eating addiction" (Hebebrand et al., 2014), but as I've stressed previously, an addictive behavior is not the same thing as an addictive substance. People who are addicted to eating might well find foods—including sugar—deeply rewarding, but it's premature to blame the sugar itself. There's no problem with saying we find sugar rewarding. Of course we do, as we might exercise and so forth. To say it's addictive makes a far stronger claim, and that claim is the suggestion that it's a sinister substance that's out to get you, rather than something you're free to enjoy in moderation, just as you can alcohol or any other drug, depending on the legislation where you live, and your own personal risk tolerance.

CONCLUSIONS

The concern expressed in this chapter is that the tone of much of what we read and hear can compromise one's risk tolerance through treating humans as perpetual victims, consuming foodstuffs that are out to kill them. But what reading the actual papers, rather than media coverage of them, mostly reveals is that science journalists no longer read or understand the journals, and that the public—and some professionals—are far too trusting when it comes to the sensational headlines that convey elements of those studies to us.

On the sugar question specifically, the point, in short, is quite simply that any claim that sugar is "addictive" is arguably using the word "addictive" in a misleading and hyperbolic way. And, as the cases and science detailed in Johann Hari's *Chasing the Scream* (2015) persuasively suggest, the primary vehicle we have for escaping addiction is to give ourselves a sense of purpose and, above all, agency—and agency is the last thing that panics around things like sugar addiction have time for. Instead, the narrative is all about you being a victim of conspiracy. This sells books and creates gurus who can "help" you out of darkness—but sometimes this comes at the cost of careful reasoning.

We should not offer people this sort of dogma, both because it dumbs down the process of scientific reasoning and because it encourages people to think in terms of false dichotomies or other poorly defined and crude categories. What's right is often nuanced, and might not fit in a tweet or a newspaper headline, and almost certainly won't sell books.

But the problem is that we like stories, and the media feed that liking. Sensationalist stories gain traction via our confirmation bias, and our cognitive dissonance—not being able to reconcile the complicated version of events with the sensationalist one—results in the backfire effect, in which we double down on our existing beliefs. What this adds up to is hysteria and moral panic, with little tolerance for nuance.

Worse of all, sometimes people who work in science contribute to our illiteracy by cherry-picking data, by presenting science as settled when it's actually contested. This might sell books and gain a following, but at the expense of our critical-thinking skills. The democratization of knowledge via the Internet has brought real boons to society. But it can also make us forget that real scientific breakthroughs happen in journals, not in bestselling cookbooks. And you hear about them on the news, not on the Dr. Oz show.

Our worth as critical thinkers is not vested in *conclusions*, but in the manner in which we reach conclusions. It's not about merely *being right*. Being right—if we are right—is the end product of a process and a method, not an excuse for some sanctimonious hectoring or dietary evangelism. As Oscar Wilde (2006) had it, "the truth is rarely pure and never simple."

QUESTIONS FOR REFLECTION

1. *On balance do you think that popular media, including social media, inform or misinform us? Are there certain outlets that are more or less likely to inform or misinform?*
2. *Which topics of conversation are most likely to lead you, personally, to engage in motivated reasoning?*
3. *Programs from Apple, Google, and Microsoft allow you to seamlessly carry information, including browser search histories, with you from your desk at home to*

the office—or even to the beach! Do you think that it's possible to avoid living in a filter bubble in this type of a hyperconnected life?

4. *As reviewed previously, even heroin isn't quite as addictive as many of us believe. How would you differentiate between heroin addiction and so-called Internet addiction?*
5. *Should scientists be more engaged with the public in order to disseminate their findings publicly? What might the downside of doing so be?*

REFERENCES

American Psychiatric Association. (2000). *Diagnostic and statistical manual of mental disorders* (4th ed., text rev; *DSM-IV-TR*). Washington, DC: American Psychiatric Association.

Avena, N. M., Rada, P., & Hoebel, B. G. (2008). Evidence for sugar addiction: Behavioral and neurochemical effects of intermittent, excessive sugar intake. *Neuroscience & Biobehavioral Reviews*, *32*(1), 20–39.

Bohannon, J. (2015). *I fooled millions into thinking chocolate helps weight loss. Here's how*. Retrieved from http://io9.com/i-fooled-millions-into-thinking-chocolate-helps-weight-1707251800

Carroll L. (1991). *Through the looking glass*. Project Gutenberg. Retrieved from http://www.gutenberg.org/files/12/12-h/12-h.htm

Fernbach, P. M., Rogers, T., Fox, C. R., & Sloman, S. A. (2013). Political extremism is supported by an illusion of understanding. *Psychological Science*, *24*(6), 939–946.

Fisher, M., Goddu, M., & Keil, F. (2015). Searching for explanations: How the Internet inflates estimates of internal knowledge. *Journal of Experimental Psychology: General*, *144*(3), 674–687.

Goldacre, B. (2014). Preventing bad reporting on health research. *British Medical Journal*, *349*, g7465.

Hall, H. (2014). Does the movie fed up make sense? *Science-Based Medicine*. Retrieved from https://www.sciencebasedmedicine.org/does-the-movie-fed-up-make-sense

Hari, J. (2015). *Chasing the scream: The first and last days of the war on drugs*. New York, NY: Bloomsbury.

Harvard HelpGuide. (2015). *Understanding addiction: How addiction hijacks the brain*. Retrieved from http://www.helpguide.org/harvard/how-addiction-hijacks-the-brain.htm

Hebebrand, J., Albayrak, O., Adan, R., Antel, J., Dieguez, C., de jong, J., . . . Dickson, S. L. (2014). "Eating addiction", rather than "food addiction", better captures addictive-like eating behavior. *Neuroscience & Biobehavioral Reviews*, *47*, 295–306.

Hyman, M. (2014). Sweet poison: How sugar, not cocaine, is one of the most addictive and dangerous substances. *New York Daily News*. Retrieved from http://www.nydailynews.com/life-style/health/white-poison-danger-sugar-beat-article-1.1605232

Langreth, R., & Stanford, D. (2011). Fatty foods addictive as cocaine in growing body of science. *Bloomberg News*. Retrieved from http://www.bloomberg.com/news/2011-11-02/fatty-foods-addictive-as-cocaine-in-growing-body-of-science.html

Martin, A. (2013, May 1). *The web's 'echo chamber' leaves us none the wiser*. Retrieved from http://www.wired.co.uk/news/archive/2013-05/1/online-stubbornness

McRaney, D. (2011). *The backfire effect*. Retrieved from http://youarenotsosmart.com/2011/06/10/the-backfire-effect

Munroe, R. (n.d.). *Headlines*. Retrieved from http://xkcd.com/1283

Nickerson, R. (1998). Confirmation bias: A ubiquitous phenomenon in many guises. *Review of General Psychology*, *2*(2), 175–220.

Nyhan, B., & Reifler, J. (2010). When corrections fail: The persistence of political misperceptions. *Political Behavior*, *32*(2), 303–330.

Oransky, I. (2015). Should the chocolate-diet sting study be retracted? And why the coverage doesn't surprise a news watchdog. *Retraction Watch*. Retrieved from http://retractionwatch.com/2015/05/28/should-the-chocolate-diet-sting-study-be-retracted-and-why-the-coverage-doesnt-surprise-a-news-watchdog

Pariser, E. (2011, March 3). *Beware online "filter bubbles."* TED Talks, Long Beach, CA. Retrieved from http://www.ted.com/talks/eli_pariser_beware_online_filter_bubbles

Pullen, J. (2015, April 20). Science says you should ignore Internet trolls. *Time Magazine*. Retrieved from http://time.com/3827683/internet-troll-research

Silverman, C. (2011). The backfire effect: More on the press's inability to debunk bad information. *Columbia Journalism Review*. Retrieved from http://www.cjr.org/behind_the_news/the_backfire_effect.php?page=all

Sumner, P., Vivian-Griffiths, S., Boivin, J., Williams, A., Venetis, C. A., Davies, A., . . . Chambers, C. D. (2014). The association between exaggeration in health related science news and academic press releases: Retrospective observational study. *British Medical Journal*, *349*, g7015.

Wilde, O. (2006). *The Importance of Being Earnest: A Trivial Comedy for Serious People*. Project Gutenberg. Retrieved from http://www.gutenberg.org/files/844/844-h/844-h.htm

CHAPTER 7

ALIENS, ABDUCTIONS, AND UFOs

Imagine that you and your significant other are driving home after a relaxing vacation. It's late at night and there are few other cars on the road. You notice some odd lights in the sky, moving erratically, and pull over to get a better look. At first you think it is a plane, but no plane can move like these lights are doing . . . and it now seems to be coming down from the sky toward you. You get back into the car and the two of you speed off down the road. The lights follow you, and you see it is now some sort of craft, at least 100 feet long and moving up and down in the sky. Eventually it descends rapidly from the sky, causing you to slam on the brakes and stop in the middle of the road. The craft hovers off the ground, filling the view from your windshield. You step out of the vehicle and see numerous humanoid shapes in the craft, looking down at you. A door opens in the bottom of the craft, and you run back to the car, afraid for your life and that of your significant other. You speed off, trying to keep an eye on the craft and its occupants, but find that it has disappeared from view. Suddenly, the car begins to shake and you feel an odd tingling throughout your body. Your watches stop working and your mind goes dull . . . and then you come to, 35 miles down the road with no idea what has happened in the past 2 hours. You keep driving and finally arrive home hours later than you should have. Two years later, under hypnosis, you are finally able to recall the events of that night, which included being taken aboard an alien craft and being given a medical examination of some kind. This scenario is the story of Betty and Barney Hill, who relayed that all this happened to them on September 19, 1961.

Or imagine jolting awake during the night, with your eyes shooting open and your heart feeling as if it's about to beat out of your chest. You are covered in sweat, it's dark, and you can't see well. You feel the presence of someone or something in your room, and catch a glimpse of a shadowy figure moving along one wall. You try to turn your head and reach for the light, but can't move. Your entire body is paralyzed, except for your eyes. Your body is in full fight-or-flight mode and you are absolutely terrified, struggling to move, to do anything to break the spell that has come over you. You feel as if your bed is vibrating slightly, you hear an odd humming, and still the menacing figures are moving around your room, staring at you.

After an unknown amount of time, you feel yourself dropping back into unconsciousness as the humming continues and the figures still watch. You wake the next morning feeling groggy and tired, unable to shake the feeling that something terrible happened during the night. This scenario happens to millions of people across the world on a monthly basis.

Finally, imagine that you are a rancher in rural New Mexico. One day while working, you notice an odd pile of debris on your property, something that looks like the remains of a flying object of some kind. Not knowing what to make of it, you contact the local authorities, and a major in the U.S. Army comes to your ranch, confiscates the materials, moves them to the nearest military base, and tells you and the local press that it was the remains of a high-altitude weather balloon. Satisfied, you don't think much of the incident for the next few decades, until you read a news article about a new book that has come out, detailing how researchers have interviewed hundreds of individuals who worked at that military installation. They concluded that, rather than some "weather balloon," the wreckage was actually the remains of a crashed alien spacecraft, including extraterrestrials, and that the government had been involved in a massive cover-up of this incident for over 30 years. Stunned, you wonder exactly what it was that crashed on your property that day, and what else they are hiding from the public. This is the story of what happened to William Brazel, a ranch foreman who worked on a homestead approximately 30 miles north of Roswell, New Mexico, during the summer of 1947.

Each of these stories reflects a different aspect of the answer to one question: Does intelligent life exist elsewhere in the universe, and has it visited Earth? This chapter explores the various scientific and pseudoscientific lines of thinking about extraterrestrial life, alien abductions, and related issues, using everything you have learned so far in this book to examine such claims.

LIFE OUTSIDE OF EARTH

Space, at least as described by Gene Roddenberry in the introduction to each episode of *Star Trek*, is the final frontier. Few things can provoke as much wonder and awe as staring into the clear night sky, far away from cities and bright lights, and seeing the moon, planets, and thousands of stars that fill it. Despite over 50 years of space science and exploration, we still know so little about the enormity of what lies beyond our atmosphere and our solar system. The Voyager space probes, launched in 1977 from Florida, took over 35 years to reach the edge of our solar system, even traveling at over 60,000 kilometers per hour! We have sent probes to all of the solar system's planets and robots to Mars, even recently landed a probe on a comet, and still have barely scratched the surface of exploring even our own nearest neighbors.

Is it any wonder that so many people over the years—writers, artists, scientists, and the general public—have been curious about the possibility of life outside of our planet?

Based on public opinion polls, a substantial number of people worldwide don't only think about aliens, but think they exist and have visited our planet. Surveys since the early 2000s showed that 30% of the general public believed that "some of the unidentified flying objects that have been reported are really space vehicles from other civilizations" (National Science Foundation, 2001) and 32% thought that unidentified flying objects (UFOs) were alien space craft (Harris Interactive, 2009). Another poll found that over 70% of people thought the U.S. government was involved in hiding or covering up knowledge about the existence of alien life visiting Earth (Newport, 1997). The highly popular television show *Ancient Aliens*, which explores the hypothesis that aliens have visited the Earth numerous times in the past and influenced the development of human civilization, has run for seven seasons and garners over 1 million viewers per episode.

Scientists have also been intensely curious about the possibility of life outside our planet for quite some time. One of the earliest attempts to spur scientific interest in life outside of Earth was by astronomer Frank Drake, who held the world's first "search for extraterrestrial intelligence" (SETI) conference in 1961. In his words (Drake, 2003),

> As I planned the meeting, I realized a few days ahead of time we needed an agenda. And so I wrote down all the things you needed to know to predict how hard it's going to be to detect extraterrestrial life. And looking at them it became pretty evident that if you multiplied all these together, you got a number, *N*, which is the number of detectable civilizations in our galaxy. This was aimed at the radio search, and not to search for primordial or primitive life forms.

The equation that Drake came up with is $N = R_* \times f_p \times n_e \times f_l \times f_i \times f_c \times L$. In it, *N* is the number of civilizations in our galaxy with which radio communication might be possible. You would find *N* by multiplying the following quantities:

- R_* = the average rate of star formation in our galaxy
- f_p = the fraction of those stars that have planets
- n_e = the average number of planets that can potentially support life per star that has planets
- f_l = the fraction of planets that could support life that actually develop life at some point
- f_i = the fraction of planets with life that actually go on to develop intelligent life (civilizations)

- f_c = the fraction of civilizations that develop a technology that releases detectable signs of their existence into space
- L = the length of time for which such civilizations release detectable signals into space

The equation was very effective in stimulating ideas and research into these factors, even though the last four are basically unknown as quantities. Using best estimates, Drake et al. concluded that the likely number of civilizations in our galaxy was between 1,000 and 100,000,000 (Drake & Sobel, 1992). Even highly conservative estimates, though, place N at far above 1 (which is the number of radio-producing civilizations known to exist . . . that's us humans, by the way).

Drake's work inspired numerous other scientists, leading to what is now known as the field of astrobiology. Astrobiology focuses on understanding how life arises in the universe, both on Earth and elsewhere, and is a highly multidisciplinary endeavor, bringing together chemists, biologists, physicists, astronomers, and more. The National Aeronautics and Space Administration (NASA) has an Astrobiology Institute that has recently focused on examining meteorites for traces of microscopic life, leading the hunt for exoplanets (planets outside our solar system), and researching extremophiles (life that exists in very inhospitable environments, such as highly acidic or extraordinarily hot places). The SETI Institute has also played a major role in the search for non-Earth life, examining the skies for radio wave communications potentially emanating from other planets and conducting research on the basic chemistry of life.

Recently, planetary scientist Sara Seager revisited the Drake equation to reflect the search for not just intelligent life, but *any* life forms on other planets. Rather than focus on radio waves, she changed her focus to examining the atmosphere of exoplanets for traces of oxygen, water vapor, and other gases that are indicative of life. Her equation was $N = N_* \times F_Q \times F_{HZ} \times F_O \times F_L \times F_S$, where N is the number of planets with detectable signs of life. That would be estimated by multiplying these factors:

- N_* = the number of stars observed
- F_Q = the fraction of stars that are quiet
- F_{HZ} = the fraction of stars with rocky planets in the habitable zone
- F_O = the fraction of those planets that can be observed
- F_L = the fraction that have life
- F_S = the fraction on which life produces a detectable signature gas

As with the Drake equation, some of these values are known and some (the last two) are guesses. Still, even with conservative estimates for the unknowns, Seager hypothesizes that we should be able to find two planets that are actively inhabited by some type of life before 2025, given the new satellites and scientific missions being launched in the next few years. This idea was given further support when, in the summer of 2015, NASA announced the discovery of "Earth 2.0," the first time a near Earth-sized planet in the habitable zone of a sun-like star had been found. These types of calculations and other lines of evidence have recently led Stephen Hawking to say, "To my mathematical brain, the numbers alone make thinking about aliens perfectly rational" (Boyle, 2010).

So the general scientific consensus appears to be that life is most likely out there, somewhere, in the galaxy. Large portions of the populace believe that aliens have visited Earth, either in the distant or recent past. A smaller but not insubstantial number of individuals (anywhere from 1,700 people to 5% of the population, depending on who you get your information from; Appelle, 1996) believe that they have personally been abducted by aliens. And yet, as we will see, there is no definitive, physical evidence of life outside our planet, let alone intelligent life that has traveled interstellar distance to visit our planet. Why is there this kind of belief–evidence gap when it comes to aliens?

ALIENS THROUGHOUT HISTORY

The concept of nonsupernatural, intelligent beings living elsewhere than on Earth is a very old one, with the earliest known reference being found in 2nd-century satirist and skeptic Lucian of Samosata's *True History*. In it, Lucian relates how he visited the moon and found it inhabited by alien beings who were at war with the inhabitants of the sun. Ancient Arabic and Japanese tales also feature planets outside our own that are inhabited, although they are few in number compared to supernatural tales featuring gods, demons, and other supernatural creatures. Writers such as Jonathan Swift and Voltaire played with the theme of visitors from other worlds, and a small number of not-very-well-known writers in the late 1880s set stories on the planet Mars or wrote about invasions by alien beings from other worlds. However, the notion of beings from another planet visiting Earth for nefarious purposes really came to the public attention with the H. G. Wells novel *The War of the Worlds*, first published serially in 1897.

Wells's story detailed the events of an invasion of Earth by creatures from Mars. A cylinder-shaped craft crashes to the ground, disgorging two Martians who are described as being gray with oily skin, having two large dark-colored eyes, a lipless mouth, and a mass of tentacles (see Figure 7.1, the cover of a 1927 *Amazing Stories* reprint). These creatures then proceed to place themselves in large war machines and terrorize the British landscape, all the while kidnapping victims for unknown purposes. The aliens are finally defeated not by man, but by nature: they are killed by exposure to Earth's bacteria.

FIGURE 7.1 War of the Worlds.
Source: Courtesy of Wikimedia, interior Illustration from the reprinting of the H. G. Wells' novel in *Amazing Stories*, 1927. Illustrated by Frank R. Paul.

The War of the Worlds was hugely popular, both when originally published serially and when later collected as a novel. It was also massively influential on the developing literary genre of science fiction, particularly because of its depiction of what aliens looked like and what their goals were. Interestingly, the "big head, big eyes, small body" look of these (and later) aliens likely owes itself to speculation in the 1890s about what humans would eventually evolve into as we came to rely less on our physical forms and more on our minds. Wells and others had hypothesized that our brains would grow increasingly large and our bodies would become smaller with disuse. Wells had even written about this idea in 1893 in *Of a Book Unwritten, The Man of the Year Million*, describing humans of the future as having a small body, enormous head, no hair, no mouth, and no nose. Interestingly, he was preceded in this notion (and likely influenced) by the 1891 Kenneth Folingsworth book *Meda: A Tale of the Future*, which describes small gray men with huge heads as being the "future" of humankind. *The War of the*

Worlds, though, was massively more popular and succeeded in helping this image of "aliens" become ensconced in our cultural consciousness.

Over the next few decades, science fiction exploded as a genre, with books, radio dramas, pulp magazines, comics, movies, and television all exploiting and adding to Wells's themes about aliens (Westfahl, 2005). Early portrayals of aliens were highly varied, from looking like humans with only a minor change, such as different skin color, to closely resembling some type of animal, but certain themes began to emerge regarding how aliens coming to Earth would act. Through stories in pulp magazines like *Amazing Stories*, comics such as *Flash Gordon*, and movies like *The Day the Earth Stood Still* and *Invasion of the Saucer-Men*, a typical type of "alien story" developed. The aliens were increasingly displayed as humanoid, but with giant eyes and heads and small bodies (see Figure 7.2). They were always technologically advanced compared to humans, and were frequently hostile. They were traveling in some sort of saucer-shaped craft and often captured people (particularly females) in remote areas to experiment on them. The people that were taken were put into trances or hypnotized in some way, resulting in periods of missing time.

BETTY AND BARNEY HILL'S ABDUCTION

By the early 1950s, there were several published accounts of individuals being abducted or seeing others abducted (see Rogerson, 1993 for a concise history). The number of supposed abduction tales began to accelerate in

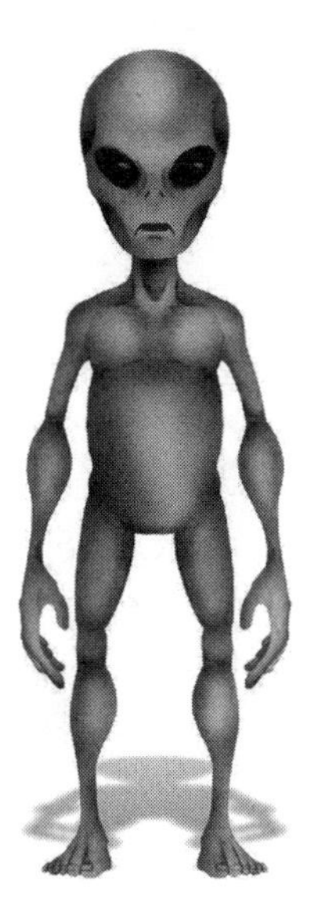

FIGURE 7.2 Gray alien.
Source: Coutresy of Wikimedia. Drawing by LeCire.

1954, with reports being found across the world, from France to Denmark, Venezuela to Brazil. The aliens described by early "abductees" were not uniform in any way, described in such varied ways as rainbow-colored shapeshifters, walking asparagus stalks, elephant-trunked men, and more. The accounts of the experience were also widely different, with some people reporting just seeing the aliens and escaping, some reporting multiple day abductions, and one early tale even reporting a 2-year abduction. Many of the reports borrowed liberally from the fictional accounts of aliens from literature and film, or appeared to blend traditional supernatural beliefs and the new "secular" alien accounts. It wasn't until Betty and Barney Hill's reported abduction in 1961 that accounts really exploded . . . and began to be eerily similar to one another.

As described at the start of this chapter, the Hills had a very odd experience one night, seeing a flying object with beings inside that appeared to chase them, and then having no recollection of 2 hours' and about 35 miles' worth of driving. Initially, that was all that Betty told her sister, who encouraged them to report the incident to the authorities. She did, calling the local Air Force base 2 days after the incident. They investigated but found no compelling evidence of anything unusual, instead saying that they had likely just misidentified a known object, like a planet. Betty, however, seems to not have accepted this and started seeking out information on UFOs, checking out a book on them from the local library called *The Flying Saucer Conspiracy* (Keyhoe, 1955). She soon wrote to the author, a retired Marine Corps officer and head of an amateur UFO research group. Her letter was eventually passed on to another UFO enthusiast and astronomer, who conducted an extensive interview with the Hills approximately 1 month after their experience.

During this time (before the interview and after reading the UFO book), Betty began having very intense, disturbing dreams. She later recalled them as being about the "missing time" that she and Barney experienced, and over a month later wrote the dreams down. Betty described how in her dream they were taken aboard a disk-shaped craft and given a series of medical-style examinations. She also said the beings in her dreams were short, with dark hair and eyes, and very large noses. Over the next 2 years, the Hills would speak to numerous people, from their church group to amateur UFOologists, about their experiences. From their own reports, Betty and Barney were quite upset by the encounter and were experiencing high levels of anxiety. It was this and the prompting of an Air Force officer who had become interested in their case that eventually led them into the office of Dr. Benjamin Simon, a noted psychiatrist and hypnotist.

Dr. Simon met with the couple for individual sessions over the next 6 months, keeping extensive records. During the sessions, Barney described being taken aboard the craft and given a medical exam, although with less detail than Betty had discussed in her dreams. He said that his intense fear made him shut his eyes during most of the experience, but also put a large emphasis on the beings' eyes, reporting them as terrible and yet feeling they were too irresistible to look away from. Betty's recollections of the experience under hypnosis were very similar to the dream content she had reported earlier, but highly fleshed out and the order of the experiences was remembered slightly differently. This included Betty remembering seeing a three-dimensional map filled with stars and various lines connecting them. She later drew this "star map" for Dr. Simon. Intriguingly, both Barney and Betty displayed very high amounts of distress, fear, and anxiety during their earlier sessions, but by the end of their work with him reported feeling at peace and not anxious about their experience. It should be noted that Dr. Simon did not think that the couple had actually been abducted, but he did acknowledge the very high distress the couple were experiencing. He thought their experience was a fantasy, the result of high psychological stress, rather than an actual abduction.

The Hills led fairly normal, quiet lives for the next year, until a front-page story in a Boston newspaper about them and their experiences was published. The story was picked up by the national and international press, leading to huge amounts of publicity around the couple. They were soon contacted by author John G. Fuller, Jr., who had written about UFOs before, and agreed to cooperate with him on a book about their experience. *The Interrupted Journey* (Fuller, 1965) was a huge success and was reprinted numerous times, with excerpts published in magazines and a positive review published in *TIME* magazine. This eventually led to a made-for-television movie called *The UFO Incident* (starring a young James Earl Jones as Barney Hill!) in 1975 and references to the Hills in numerous television series, including Carl Sagan's *Cosmos*, *The X-Files*, and *Taken*. More recent, a book by Betty's niece Kathleen Marden and UFO researcher Stanton Friedman called *Captured! The Betty and Barney Hill UFO Experience: The True Story of the World's First Documented Alien Abduction* (2007) expanded on known accounts by including some transcripts of the hypnosis sessions. Although Barney died unexpectedly from a stroke in 1969, Betty became a major celebrity in UFO enthusiast circles and spoke frequently about the experience until her death at age 85 in 2004.

CRITICAL THINKING ABOUT ABDUCTION CLAIMS

The story of Betty and Barney Hill is both an archetypal story of an other-worldly encounter and the prototypical story of an alien abduction. It is also an excellent way to exercise the critical-thinking skills you began developing in the first half of this book. By examining the Hill story (and, as we will see, many other claims) through a critical-thinking lens, rather than just taking it at face value, certain issues arise with the story of a couple who were abducted by aliens on a lonely highway. As discussed in Chapter 4, certain aspects of skeptical and scientific thinking can be distilled into some basic critical-thinking tools we can use. By choosing only two of these, we are able to poke some fairly large holes in this story's cloth.

First, the claims of the Hills—that they were taken aboard an alien spacecraft, where medical tests were done on them, and then had their memories mostly removed—is quite extraordinary. So does the evidence support such a claim? There was, for example, no conclusive physical evidence produced to support the claims. The dress that Betty was wearing that night was actually subjected to a number of chemical analyses over the past few decades, none of which found any definitive evidence of extraterrestrial matter on it. The Hills did not have any physical reminders of their experience—no scars from where a probe was allegedly shoved into Betty's belly button, for example. Although some proponents of the abduction claim that there is copious physical evidence, including Barney's scuffed shoes and Betty's torn dress, there is no proof that either were caused by some sort of alien abduction. The only evidence we have is the testimony of the Hills prior to, during, and after their hypnosis sessions. Given this, we have to pause and take stock of what we know about how human memory works, and how it interacts with hypnosis.

Memory is a funny thing. Rather than just recording everything that we see, feel, and do similar to the way a video camera works, memory is primarily reconstructive. This means that we do not simply "play back" our memories, experiencing them in the exact same way each time. Instead, we quite literally recreate our memories each time we bring them to mind, so how accurate they are is therefore subject to many, many variables. For instance, our past experiences, current emotional state, and even social situation can all change what information we are able to recall about events. Shermer (2002) described memory as "a complex phenomenon involving distortions, deletions, additions, and sometimes complete fabrication" (p. 96). This is not to say that we just "make up" our memories, but instead that we can (and often do) remember events and experiences in ways that are not actually true. Here is a quick example: Close your eyes and think about the last time you went shopping. Visualize what that trip was like for about 10 seconds and then open your eyes again. And . . . go!

Now, did you see yourself in your memory, from a third-person perspective, or did your memory consist of a first-person view, as you actually see the world? Chances are you saw yourself moving through the store, picking up items as you went . . . which is not what you actually saw in real life! If memory worked as most people assume it does, then you would have only remembered shopping through what you saw with your own eyes, from a first-person perspective. Instead, you reconstructed the memory, putting together pieces of various shopping experiences as well as what you look like (or think you look like) and recalling that as accurate. We know from decades of research conducted by psychologists such as Elizabeth Loftus and Ulric Neisser (among many, many others) that eyewitness memory is highly fallible, and we even know the types of circumstances that can make memory even more unreliable.

For instance, Loftus's work has shown that the way questions are phrased can have a major impact on what type of information is recalled. For example, asking whether a car "hit" or "smashed" into a stop sign can cause people to change their estimates of how fast a car is moving in a video (Loftus & Palmer, 1974). Loftus also was critical in examining how false memories could be easily inserted into someone's mind, demonstrating under highly controlled circumstances that she could cause people to recall entirely fictional events as if they had truly occurred, such as being lost in the mall as a child (Loftus & Pickrell, 1995). Neisser's work repeatedly demonstrated how we misremember even highly traumatic and impactful events that seemed to be burned into our memory (so-called "flashbulb" memories; e.g., Neisser & Harsch, 1992). Even President George W. Bush, when asked how he learned about the 9/11 attacks, gave three separate and conflicting accounts within just 4 months of the event (Greenburg, 2006).

One of the major factors that can impact memory is stress. Being under high amounts of stress and anxiety can cause problems with both the encoding of information (that is, putting the memory in) and the retrieval of information (i.e., taking the memory back out), even in otherwise healthy individuals (Schwabe, Joels, Roozendaal, Wolf, & Oitl, 2012). Another typical cause of confabulation (the technical term for remembering something that did not actually happen) is co-witness contamination. This is the name for when two or more people experience the same event and they then discuss it with each other. They will often integrate details from each other's account and recollection into their own account, creating false memories about the event. Again, this does not occur purposefully or maliciously, but is a natural by-product of how our memories work—by reconstructing information, not just wholly recollecting it.

So memory, it turns out, is highly fallible, especially in times of high stress, and discussing the event with other eyewitnesses can cause even more errors to appear. This is true even of events of high impact or trauma that seem to be burned into our memories. But what of hypnosis? How can it help or harm the way you remember things?

First, there is widespread belief among both laypersons and clinicians that hypnosis can enhance one's memory, even to the point of being able to recall memories of traumatic events that someone does not consciously have access to (Patihis, Ho, Tingen, Lilienfeld, & Loftus, 2014). However, careful research over the past few decades has repeatedly shown that hypnosis actually makes memory recall *less* accurate, not more. Rather than helping to break down barriers to recall, being under hypnosis makes people recall larger amounts of information than they normally would, but this information is more prone to errors and as such helps in the creation of false memories (Mazzoni & Lynn, 2007).

In addition to these problems with memory, it turns out that humans are spectacularly bad at recognizing and identifying objects in the night sky. This is frequently caused by misidentification or misperception of well-known objects. For example, planets (especially Venus), meteors, man-made objects (such as aircrafts or satellites), and even weather phenomena have been found to decisively account for the vast majority of reported UFO sightings, from 69% by the United States Air Force (Ruppelt, 2011) to 89% by the Center for UFO Studies (Hendry, 1979).[1]

To add to the confusion of trying to figure out what one is seeing, looking at small, bright objects against a dark background (as happens when seeing stars, planets, airplanes, and such at night) causes a specific perceptual effect dubbed the autokinetic effect. When this occurs, the stationary points of light appear to move, sometimes changing direction back and forth. This happens because your brain is attempting to control for the natural random drift of your eyes but cannot, because there's no frame of reference present. You can try this yourself at home by taking a toilet paper roll, walking outside on a starry night, finding a star, putting the roll up to your eye, and staring at the star. Soon, it will appear to be swaying or moving, even though it's stationary. This can cause people in moving vehicles to think that an object is chasing them or moving away from them, as in the very well-documented cases of police officers in both Ohio in 1966 and Devon, England, in 1967 . . . who were actually found to be "chasing" the planet Venus at speeds of 90 miles per hour (Sheaffer, 1986). When this is added to the fact that human perception is terrible at judging distances or speed of lights in the sky, it adds up to a lot of normal, recognizable objects being called UFOs. On this point, noted science fiction author and skeptic Arthur C. Clarke (of *2001* fame) once said, "If you've never seen a UFO, you're not very observant. And if you've seen as many as I have, you won't believe in them."

With both the problems of memory and the problems with identifying objects in the night sky well known and studied, one must consider the situation surrounding the Hills and their extraordinary claims in a new light. Our understanding of the world as it is (including how memory works, the impact of hypnosis on memory, and how perception is easily fooled) casts serious doubt on the accuracy of the Hills' eyewitness testimony. Specifically, Barney was known to be under high amounts of stress during the time of the encounter, a major reason they went away for a few days of rest and relaxation. They were also driving back very late at night and it is likely that

[1] For a more in-depth look at explainable astronomical causes of UFOs, please see Ridpath (2012).

they were physically tired. The road they were on was not well traveled, especially late at night, and would have had a spectacularly dark sky during that time. All aspects of their initial story (seeing a bright object moving in the sky, and then it chasing them and supposedly getting closer) are easily explainable in natural, nonextraterrestrial terms. The Hills then talked to one another and numerous other people about their encounter, including reading books on aliens and UFOs, during the 2 *years* between the incident and their hypnosis sessions. This and the fact that their stories got more and more similar after hearing each other's hypnosis-induced recall is a classic example of co-witness contamination.

As this analysis demonstrates, not only is the best evidence for the Hills' claims truly unextraordinary, but it also fails a number of other critical-thinking tests. For example, there is quite literally no way to test their claim in a falsifiable manner, given that it is purely based on their testimony. Applying Occam's razor shows that the most parsimonious, least assumption-filled explanation is that the Hills misidentified a heavenly body (likely Jupiter, according to initial and subsequent analyses) and then developed a series of false memories surrounding the events thanks to the highly malleable nature of human memory.

This explanation leads into one of the many concerns that skeptics have had about Betty's testimony in particular. She, from at least the mid-1960s until her death, reported seeing alien spacecraft everywhere she went. In a commonly quoted letter from 1966, she wrote to a friend that "Barney and I go out frequently at night for one reason or another. Since last October, we have seen our 'friends' on the average of eight or nine times out of every ten trips" (as quoted in Friedman & Marden, 2007, p. 208). During talks she gave and in books she wrote following her rise to UFO stardom and after Barney's death, she described seeing trucks levitating as a result of UFOs, large numbers of UFOs flying together, watching aliens do physical exercise upon leaving their craft, and many other stories that further strain her credibility. When one sympathetic UFO researcher accompanied Betty to her favorite reported UFO "landing spot," he reported that she was "unable to distinguish between a landed UFO and a streetlight" (Burke, 1977). This is, to say the least, pretty damaging to the potential accuracy of her initial testimony.

One aspect of the case that many point to as being strong evidence for the reality of the experience is the high amounts of reported distress and fear that both Betty and Barney displayed during the hypnosis sessions. Why, if these memories were not actually real, would one be so upset? And, why did they both report feeling so much better after the hypnosis sessions? Intriguingly, psychologist Richard McNally conducted extensive research that helps to answer this question (see McNally [2012] for an in-depth summary of this work). In a series of studies in the first decade of the 21st century, McNally and his colleagues helped shed light on the psychological and physiological characteristics of those who claim to be abducted. In one particularly salient study, they took measures of autonomic arousal (e.g., heart rate, galvanic skin response, muscle tension) from supposed abductees who heard a narrator read their previously written-down accounts of their abduction. Amazingly, the physiological response of these abductees was at the same level as that of Vietnam War veterans who heard scripts of their war traumas! Further work showed that, compared to a control group, "abductees were neither depressed nor anxious, but they had reported unusual alterations in consciousness, belief in unconventional modes of causation, and had vivid imaginations and a rich fantasy life" (McNally, 2012, p. 8).

Based on a large body of research, one of his students, Susan Clancy, wrote an entire book looking into what the "formula" was for becoming an abductee. She identified five common factors that distinguished those who believed they had been abducted from the general population (Clancy, 2007). First is a very high level of what is called "magical thinking." That is, they endorse a large number of pseudoscientific and paranormal beliefs (such as believing in ghosts, astrology, precognition, and more). As such, the abductees appear to be very open to ideas that are supported by little to no evidence, and in fact are not concerned with the type of evidence that scientists often demand.

Second, their supposed alien encounters almost all started out with episodes of sleep paralysis and hypnopompic hallucinations. If you flip back to the start of this chapter and reread the first paragraph, it is an accurate description of this combination of events. We know it's accurate because it has been happening to your first author (CWL) for over 30 years now. Sleep paralysis results in you waking up suddenly but being unable to move your body. Hypnopompic hallucinations are those that occur as you are waking up. Both are well-documented natural phenomena, even if exactly *why* they occur is not (Adler, 2010). I (CWL) have personal experience with these, having had this combination of things happen to me multiple times monthly for as far back as I can remember. I don't always see menacing human figures, but have also seen bears (in my hallway), felt snakes crawling across my face, and heard sounds like crying or screaming. If you've never had this happen to you, I honestly hope it never does. It is a truly frightening experience, even though I understand that what is happening is not reality based, but instead being constructed by my mind. I often describe the experience as having one part of my brain saying "Hey, it's okay, it's just sleep paralysis and hallucinations, you will be fine" while the other part of my brain says "AAAAAAAAAHHHHHHHHH!!!" These experiences have been reported cross-culturally throughout human history, and are most commonly attributed to some type of evil presence, such as demons, spirits, or witches.

The third common thread is that the vast majority of abductees undertook hypnotic regression sessions, not unlike what the Hills did with Dr. Simon.

The key difference, though, is that today's abductees are often able to find someone who "specializes" in alien abductions to do their sessions, someone who already believes that aliens are visiting Earth. These sessions, then, are typically filled with leading questions, and the combination of those plus hypnosis is a powerful engine for creating false memories.

The fourth factor was being very high in "absorption," a psychological trait that makes one more prone to fantasy, experiencing highly vivid mental imagery, being suggestible, and being easily hypnotizable. Finally, people must be familiar with the typical cultural narrative of what alien abduction "looks like." In other words, the abductees report being very familiar with how books, television, and movies portray both the look of aliens and the experiences of those who are supposedly abducted. In fact, as discussed earlier in this chapter, prior to the widespread publication of the Hills' story, there were numerous precursor fictional tales that set the stage for what aliens looked like and how they behaved. With the widespread publicity that Betty and Barney received, though, this crystallized into a definitive narrative of what happens during abductions—the bright light from the sky, the abduction at night in a remote area, a series of medical/sexual exams, missing time, telepathic communication, and so on. Today, very few abductees stray far from the script established in the Hill' story, although there have been increasingly fantastical elements added through the decades (such as seeing alien–human hybrids aboard the alien craft).

Intriguingly, the known facts of Betty and Barney Hill's purported abduction appear to fit quite well into this "recipe" of abduction. According to multiple sources, the Hills were both psychologically healthy (if perhaps under quite a bit of stress around the time of their experience), just as was found in McNally's work. We have already discussed the hypnosis sessions, and how Betty didn't have any memories of abduction until after she read a book on that subject and was exposed to that cultural narrative. In fact, she first had intense, vivid dreams of the abduction, not memories, *after* reading the UFO book. We also know that Betty, at least, held beliefs in paranormal phenomena such as telepathy, ghosts, and even Bigfoot (Marden, 2013); this is a clear example of the "magical thinking" identified by Clancy. As for absorption, this seems quite likely, given how fully Betty plunged into the world of UFOlogy and with her reported almost nightly sightings of UFOs. Although we don't know if perhaps either of the Hills ever experienced sleep paralysis or hypnopompic hallucinations (before, during, or after their experience), based on all available data it seems a likely event.

Despite these massive problems with the story of Betty and Barney Hill's alleged abduction (and, it turns out, every other purported abduction), this case is still held up by numerous believers in alien abductions as the best and most credible story of its kind. Indeed, many of the criticisms noted previously are dismissed out of hand by believers, including the authors of the most recent exhaustive book on the topic (Friedman & Marden, 2007). For instance, Kathleen Marden (Betty Hill's niece and purported expert on alien abductions) repeatedly points to the memories recalled by the Hills under hypnosis as proof of the validity of their claims, without discussing the problems inherent in relying on such "evidence." She has even written a defense of Betty as not being "fantasy prone" but then goes on to detail how she was persuaded to believe in ghosts and Bigfoot by a series of "paranormal researchers" (Marden, 2013). This serves as a strong reminder of how powerful our beliefs become, overshadowing our ability to objectively evaluate claims and instead ignore or twist information to fit our beliefs. As we've already seen earlier in this book (and will see many more times), you can't always trust your brain.

ANCIENT ALIENS AND MODERN CRASHES

But, even if we don't have strong evidence for people in modern times being abducted by aliens, what about other types of alien visitation? For instance, have alien spacecraft ever crashed here on Earth? Did aliens assist humans in constructing some of the greatest monuments on the planet, such as the pyramids in Egypt and Mesoamerica? Chances are you have been exposed to stories from television, the movies, or friends about one or both of these possibilities. For example, at the beginning of this chapter you read the story of William Brazel, who found wreckage that many claim is from an alien spacecraft that crashed outside of Roswell, New Mexico, and was subsequently covered up by the government. Some have gone so far as to claim that the bodies of dead alien pilots were even recovered. This event has served as the basis for numerous television shows (*Roswell*, *The X-Files*), movies (*Independence Day*, *Paul*), and books over the past 30 years. As mentioned before, one of the most popular television programs on the History Channel (reaching tens of millions of viewers per month) is *Ancient Aliens*, which purports to use archaeological and historical analyses to demonstrate how human culture and development has been extensively shaped by contact with extraterrestrials. But what does the evidence say?

As with claims of alien abduction, the stories of Roswell's crash and *Ancient Aliens* fall far short of being convincing from a scientific standpoint. Although beyond the scope of this chapter, the claims of governmental conspiracies over a crashed UFO and ancient visitation from aliens have both been thoroughly examined and found wanting. In an extensive documentary titled *Ancient Aliens Debunked*, Chris White finds that the majority of cited "examples" of ancient alien intervention in human history boils down to the logical fallacy called the *argument from ignorance*. In other words, the producers and guests are saying "Well, because I can't think of an explanation for why this looks this way, or how our ancestors could have done this with the technology available to them, then it must be aliens!" Some examples of this are how large structures, such as the Egyptian pyramids, were built or why certain figures in ancient pre-Columbian civilizations look similar to modern jets. Each, though, has a very reasonable, prosaic explanation that requires no alien intervention (White, 2013).

The Roswell incident, being much more recent in time, is even more explainable and has many parallels to the Hills' abduction story. In this case, though, there was a gap of over 30 years between the original incident and most of the initial interviews of supposed eyewitnesses, compared to the Hills. Given the lack of physical evidence for any alien craft, the more reasonable explanation for what actually crashed (it appears to have been a crashed, top-secret spy balloon from the U.S. government's Project Mogul; Korff, 1997), and the massive inconsistencies between stories told by the eyewitnesses, only someone who was falling prey to the confirmation bias could still believe that what occurred in Roswell was an alien crash. And that doesn't even begin to cover the outright hoaxes surrounding the Roswell incident, with perhaps the biggest being the "alien autopsy" video by Ray Santilli. He sold broadcast rights to this 17-minute film for millions of dollars in the mid-1990s, and it was seen by over 20 million people in the United States alone after being shown on the Fox network multiple times (Kuczynski & Carter, 2000). It wasn't until 2006 that Santilli admitted the footage was not from 1947 and that he had arranged for it to be filmed. He did, however, insist that it was a "recreation" of actual footage from a true alien autopsy, saying that he had seen the original film but that it had degraded to the point of being unwatchable. That certainly sounds plausible—right, critical thinkers?

CONCLUSIONS

Alien abductions, alien crashes, alien autopsies, and ancient aliens. Although each can be the source of fantastically entertaining fiction, all fail to pass the test of critical-thinking when examined closely. Luckily, there is an enormous wealth of scientific research involving space and potential life outside of our planet to learn about for those feeling like there is an alien-shaped hole in their beliefs about reality. For the authors, that is where the real excitement lies—not in making up stories about how we wish the universe was, but instead letting our curiosity take over and help us find out what is actually true about it.

QUESTIONS FOR REFLECTION

1. *Have you ever had an experience that you can't explain? If so, do you think it cannot be explained, or simply that you haven't found the explanation yet?*

2. *On balance of probabilities, do you think that a belief in extraterrestrial life is a responsible belief to hold? How is such a belief similar to or different from believing aliens have visited Earth, either in the past or currently?*
3. *If you were in the unfortunate position of being on trial for a serious crime that you did not commit, would you welcome the presence of someone who claims to be an eyewitness?*
4. *In terms of credibility, do you think religious claims are comparable to claims regarding extraterrestrials? How are they similar and different?*
5. *How do you account for the fascination humans seem to have with the unexplained? Is it really about aliens, or do you think it serves as proxy for something else?*

REFERENCES

Adler, S. R. (2010). *Sleep paralysis: Night-mares, nocebos, and the mind–body connection.* New Brunswick, NJ: Rutgers University Press.

Appelle, S. (1995/1996). The abduction experience: A critical evaluation of theory and evidence. *Journal of UFO Studies, 6*, 29–78.

Boyle, R. (2010, April 26). *Aliens exist, and we should avoid them at all costs, says Stephen Hawking.* Retrieved from http://www.popsci.com/science/article/2010-04/hawking-aliens-are-out-there-and-want-our-resources

Burke, G. (1977, October 15). UFO investigator refutes Betty Hill's recent claims. *Foster's Daily Democrat*. Dover, New Hampshire.

Clancy, S. A. (2007). *Abducted: How people come to believe they were kidnapped by aliens.* Cambridge, MA: Harvard University Press.

Drake, F. (2003). *The Drake equation revisited: Part I.* Retrieved from http://www.astrobio.net/topic/deep-space/alien-life/the-drake-equation-revisited-part-i

Drake, F., & Sobel, D. (1992). *Is anyone out there? The scientific search for extraterrestrial intelligence.* New York, NY: Delacorte Press.

Friedman, S. T., & Marden, K. (2007). *Captured! The Betty and Barney Hill UFO experience: The true story of the world's first documented alien abduction.* Wayne, NJ: New Page Books.

Fuller, J. G. (1965). *The interrupted journey: Two lost hours "aboard a flying saucer."* New York, NY: Dial Press.

Greenburg, D. (2006). Flashbulb memories: How psychological research shows that our most powerful memories may be untrustworthy. *Skeptic, 11*(3), 74–81.

Harris Interactive. (2009). *What people do and do not believe in.* Retrieved from http://www.harrisinteractive.com/vault/Harris_Poll_2009_12_15.pdf

Hendry, A. (1979). *The UFO handbook: A guide to investigating, evaluating, and reporting UFO sightings.* New York, NY: Doubleday.

Keyhoe, D. (1955). *The flying saucer conspiracy.* New York, NY: Holt.

Korff, K. K. (1997). *The Roswell UFO crash: What they don't want you to know.* Buffalo, NY: Prometheus Books.

Kuczynski, A., & Carter, B. (2000, February 26). Fox's point man for perversity: "World's scariest programmer,' starring Mike Darnell as himself. *The New York Times.* Retrieved from www.nytimes.com/2000/02/26/business/fox-s-point-man-for-perversity-world-s-scariest-programmer-starring-mike-darnell.html?pagewanted=all

Loftus, E. F., & Palmer, J. C. (1974). Reconstruction of automobile destruction: An example of the interaction between language and memory. *Journal of Verbal Learning and Verbal Behavior*, *13*(5), 585–589.

Loftus, E. F., & Pickrell, J. E. (1995). The formation of false memories. *Psychiatric Annals*, *25*, 720–725.

Marden, K. (2013). *Was Betty Hill fantasy prone?* Retrieved from http://www.kathleen-marden.com/was-betty-hill-fantasy-prone.php

Mazzoni, G., & Lynn, S. J. (2007). Using hypnosis in eyewitness memory: Past and current issues. In M. P. Toglia, J. D. Read, D. F. Ross, & R. C. L. Lindsay (Eds.), *The handbook of eyewitness psychology: Memory for events* (Vol. 1, pp. 321–338). Mahwah, NJ: Lawrence Erlbaum.

McNally, R. J. (2012). Explaining "memories" of space alien abduction and past lives: An experimental psychopathology approach. *Journal of Experimental Psychopathology*, *3*(1), 2–16.

National Science Foundation. (2001). *NSF survey of public attitudes toward and understanding of science and technology*. Retrieved from http://www.nsf.gov/statistics/seind06/pdf/c07.pdf

Neisser, U., & Harsch, N. (1992). Phantom flashbulbs: False recollections of hearing the news about challenger. In E. Winograd & U. Neisser (Eds.), *Affect and accuracy in recall: Studies of "flashbulb" memories* (pp. 9–31). New York, NY: Cambridge University Press.

Newport, F. (1997). *What if government really listened to people?* Retrieved from http://www.gallup.com/poll/4594/what-government-really-listened-people.aspx

Patihis, L., Ho, L. Y., Tingen, I. W., Lilienfeld, S. O., & Loftus, E. F. (2014). Are the "memory wars" over? A scientist-practitioner gap in beliefs about repressed memory. *Psychological Science*, *25*(2), 519–530.

Ridpath, I. (2012). *Astronomical causes of UFOs*. Retrieved from http://www.ianridpath.com/ufo/astroufo1.htm

Rogerson, R. (1993). Fairyland's hunters: Notes towards a revisionist history of abductions: Part one. *Magonia*, *46*. Retrieved from http://magoniamagazine.blogspot.co.uk/2013/11/notes-towards-revisionist-history-of.html

Ruppelt, E. J. (2011). *The report on unidentified flying objects: The original 1956 edition*. New York, NY: Cosimo Classics.

Schwabe, L., Joels, M., Roozendaal, B., Wolf, O. T., & Oitl, M. S. (2012). Stress effects on memory: An update and integration. *Neuroscience & Biobehavioral Reviews*, *36*(7), 1740–1749.

Sheaffer, R. (1986). *The UFO verdict: Examining the evidence*. Amherst, NY: Prometheus Books.

Shermer, M. (2002). *Why people believe weird things: Pseudoscience, superstition, and other confusions of our time*. New York, NY: Holt.

Westfahl, G. (2005). *The Greenwood encyclopedia of science fiction and fantasy: Themes, works, and wonders*. Santa Barbara, CA: Greenwood Publishing Group.

White, C. (2013). *Ancient alien evidence examined*. *Skeptic*, *18*(4), 16–23.

CHAPTER 8

PSYCHIC POWERS AND TALKING TO THE DEAD

The phone rings, and even before you look at the caller ID you know who it is. You plan to go out with some friends but decide at the last minute to stay in and study for class instead—and then you end up having a pop quiz the next day. You are talking to someone you have never met, but just know what she is going to say next. Even though your significant other is telling you that everything is fine, you sense that he is hiding his true feelings.

If you are like most people, you have had at least one of these things happen to you recently (and maybe even a few of them). But did you know that, at least according to the material on the Universal Psychic Guild's website, this means you have psychic abilities that can be unlocked with the proper training? That's right, you can learn how to unleash the hidden potential of your mind and be able to read the minds of others, see the future, and more!

Before we start looking at the evidence for and against the existence of psychic powers, though, let's take a step back and operationally define what we are going to cover in this chapter. Although the word "psychic" itself is relatively young, dating back only to the mid-1800s or so, tales of people having extraordinary mental powers date back thousands of years. Some classical examples are stories about individuals who were able to predict the future, such as the oracles and sibyls of ancient Greece; or those who communicate with the dead, such as the Witch of Endor in the Old Testament. Those types of powers and many others are usually lumped together by both skeptics and believers under the broad umbrella term of *psi* (derived from the 23rd letter of the Greek alphabet that begins the word ψυχή—psyche, mind, or soul). The Parapsychological Association (PA) uses psi to "refer to all kinds of psychic phenomena, experiences, or events that seem to be related to the psyche, or mind, and which cannot be explained by established physical principles" (Varvoglis, n.d.). This includes various forms of extrasensory perception (ESP), psychokinesis, and after-death communication (ADC).

TYPES OF PSYCHIC POWERS

Different types of phenomena are often grouped together as ESP, with the most common being telepathy and clairvoyance. Telepathy is the phenomenon usually described as transmitting some type of information directly from one person's mind to another with no physical communication between them. This can be accomplished through the insertion of thoughts (i.e., you are now hearing my voice in your head) or the interception of thoughts (i.e., I am now reading your thoughts directly from your mind). One of the most famous examples of this power would be the fictional professor Charles Xavier, leader of the X-Men, as depicted by Sir Patrick Stewart and James McAvoy in the films of the past 15 years.

Clairvoyance typically refers to being able to "see" people or events that are not physically or temporally present. There are three types generally claimed: precognition, retrocognition, and remote viewing. Precognition is the ability to foresee the future and know what will happen before it actually occurs. One of the most famous examples of this phenomenon was seen in the 16th-century Frenchman Michel de Nostredame, more commonly referred to as Nostradamus. Nostradamus is primarily known not for his early career as an apothecary, mixing herbal remedies for various illnesses, but for his later career as a seer and astrologer who wrote numerous prophecies in the 1550s that supposedly predicted events from that time to the present day, including the death of King Henry II, the rise of Napoleon Bonaparte, the world wars of the 20th century, and the Al Qaeda attacks of 9/11 (Lemesurier, 2010; Nickell, 2010b).

Retrocognition, in contrast, refers to knowing and seeing events that happened in the past (without, you know, consulting a history book or anything). For example, famous 20th-century American psychic Edgar Cayce claimed to have described the living situations and details of a minor religious sect called the Essenes in the first centuries of the Common Era (Edgar Cayce's Association for Research and Enlightenment, n.d.). He reportedly did this over a decade before the discovery of the Dead Sea Scrolls, which helped to improve our understanding of the Essenes immensely.

The final clairvoyant power is called remote viewing. It describes the ability to receive information about distant or hidden objects and events using none of your physical senses. This could be telling what objects are in a locked box, or "spying" on events many miles away using only the power of your mind. The U.S. military actually funded research, to the tune of about $20 million from the 1970s through the 1990s, into the possibility of remote viewing (this was the basis for the book *The Men Who Stare at Goats* [Ronson, 2004], later made into a 2009 movie starring George Clooney). The research was led by two trained physicists, Russell Targ and Harold

Puthoff, and employed numerous alleged psychics in an attempt to "spy" on governmental targets in the Middle East, Africa, and the Soviet Union.

In contrast to the types of ESP already described, psychokinesis or telekinesis involves the manipulation of physical objects using only the power of your mind. This could be moving small objects, such as causing dice to fall in a particular way, or causing one's self to levitate off the ground. One of the most famous psychics of the last 40 years, Uri Geller, claimed to be able to bend metal objects (he was particularly fond of spoons for some reason) and stop watch hands from moving with the power of his mind. Indian yogis have reportedly demonstrated this power for hundreds of years by levitating themselves into the air, using only a small cane for help with balance.

The last psychic power we describe before we get into our examination of such claims is a very old one, but has recently been given a very new name—ADC. In the past, ADC was known by the term "mediumship," with a medium being the person who helps those who are living communicate with those who are dead. Stories of talking to the dead are sprinkled throughout human history, but with the rise of the spiritualism movement in the United States there was a major resurgence in the idea during the late 1800s (O'Keeffe & Wiseman, 2005). Recently, many high-profile individuals, notably John Edward of SyFy's *Crossing Over* and Theresa Caputo of TLC's *Long Island Medium*, have reported being able to communicate with those who have already died, and have become quite wealthy as a result.

As related previously, we have stories, some thousands of years old, of people performing amazing feats using their minds—communicating with the dead, reading minds, seeing the future, and more. Large-scale polls among Americans and Britons indicate that substantial numbers of the public also believe these feats are real, with 41% to 50% believing in ESP and 15% to 21% in ADC (Shermer, 2011). What happens, though, once we start applying the tools of critical thinking to such claims? Do they hold up under close scrutiny, or do they collapse like Rome when confronted with the Visigoths?

It is interesting to note that there is a long history of investigation into the abilities of people claiming psychic powers, from both scientific and nonscientific standpoints. One of the earliest recorded examples of skeptical inquiry comes from the second century via the actions and writings of Lucian of Samosata, a Roman satirist (Loxton, 2012). Lucian wrote a detailed account of his encounters with a priest named "Alexander, the impostor of Abonoteichus, including all his clever schemes, bold emprises, and sleights of hand" (p. 65). In his writings, Lucian details how Alexander used many of the techniques we encounter later in this chapter to convince people that he was in contact with the gods and was able to perform "amazing" mental feats, such as being able to detail the contents of sealed scrolls and make highly accurate predictions about the future.

In the past century, well-known magicians such as Harry Houdini and James "The Amazing" Randi led a number of undercover operations to expose psychics of various kinds as fraudulent and have offered enormous sums of money to anyone able to prove that they have paranormal abilities. We have also had fairly large number of scientific studies conducted over the past 100 years to examine the possibility of psychic powers, from the early studies of botanist J.B. Rhine to experimental psychologist Ray Hyman to more recent work by eminent social psychologist Daryl Bem. As a result of the work of these and many others, a robust and compelling amount of evidence exists today to help us answer this question: Are psychic powers real?

PSI RESEARCH AND THE RISE OF PARAPSYCHOLOGY

The rise of the Spiritualism movement in the United States, England, and elsewhere during the mid-1800s gave host to a massive increase in the number of individuals who claimed extraordinary psychic powers, particularly the ability to speak to the spirits of the dead. Similarly, there was a rise in the use of scientific methodology to study and explain all types of phenomena, from the evolution of new species in biology to the inner workings of the human mind in the new field of psychology. This resulted in the turning of the scientific gaze toward paranormal claims of all kinds, whether "new age" or traditional religious ones. It was against this historical backdrop that we have the first group formed specifically to investigate such phenomena from a scientific point of view—the Society for Psychical Research (SPR), formed in London in 1882. The purpose of the SPR (as quoted in Gauld, 1968) was to be an entity that examined:

> That large body of debatable phenomena designated by such terms as mesmeric, psychical and "spiritualistic," . . . in the same spirit of exact and unimpassioned enquiry which has enabled Science to solve so many problems. (p. 137)

The SPR quickly gathered an impressive roster of academics and scientists to its ranks, began publishing a journal in 1884 (the *Journal of the Society for Psychical Research*), and quickly inspired others around the globe to take up the cause as well. For instance, the American Society for Psychical Research was formed in 1885, with a splinter group in Boston forming in the 1920s. The *Journal of Parapsychology* began publication in 1937, followed by the international Parapsychological Association in 1957. The forming of the SPR and other like-minded groups that had dedicated journals for this type of work caused a surge in the scientific research concerning psychic powers and its dissemination to a wider audience.

One of the most prominent early researchers, J.B. Rhine, would appear an unlikely candidate to study psychic phenomena and champion their existence. After receiving his master's and doctorate in botany from the University of Chicago in the mid-1920s, Rhine consequently switched fields and studied psychology at Harvard for a year before moving with his mentor, William McDougal, to Duke University. This massive change in Rhine's chosen area of study was apparently inspired by a lecture that he had attended given by Sir Arthur Conan Doyle (the creator of *Sherlock Holmes* and a vocal believer in the paranormal). While at Duke, Rhine and McDougal created the term "parapsychology" and founded the Duke Parapsychology Laboratory, a primary producer of psi-related research during the mid-20th century. It was Rhine's early experimental research into psychic powers that really set the stage for how psi-research would be conducted for the rest of the 1900s.

Given his scientific training and mind-set, Rhine wanted to carefully examine psychic powers to be certain of their validity. He even published an exposé of two famous mediums, Margery and Mina Crandon, detailing how they were just using trickery during their purported séances (Rhine, 1927, 1934).[1] This work subsequently earned him the wrath of the man who had inspired him to study psychics in the first place! In an article in a Boston newspaper, Doyle left no doubt as to what he thought about the uncovering of the fraud, writing that "J. B. Rhine is an Ass" (Polidoro, 2001).

The most well known and often cited of Rhine's works were also his earliest. In a series of experiments in the 1930s, Rhine used various protocols to test psi. One of his colleagues, Karl Zener, developed a standard deck of cards (see Figure 8.1) that could be used in various ways to test different kinds of psi. For example, to test precognition, the cards would be shuffled and laid out one at a time, face down, while the test subject would give her best guess as to which card would be laid down next. Answers would be recorded and then later compared to the actual card order, to see whether someone scored significantly higher than chance. To test for telepathy, the experimenter would again shuffle the cards, but this time would look at each card, concentrating on the image, before laying it down. The test subject

FIGURE 8.1 Zener cards.

[1] Ironically, Harry Houdini had also exposed Margery and had Doyle's anger turned upon him a few years earlier, as we see later in this chapter.

would attempt to "read" the experimenter's mind to figure out which card it was. As before, the answers would be recorded and then compared to the actual card order to look for higher-than-chance levels of correct answers, or hits.

In his first book, appropriately titled *Extra-Sensory Perception* (1934), Rhine detailed a series of experiments using Zener cards that purported to show several remarkable individuals consistently scoring very high, far above chance levels, of correct hits for both telepathic and precognitive abilities. A second book, detailing not only the results from his work but also all other published work on psi powers, followed a few years later (Rhine, Pratt, & Steward, 1940). It again concluded that the bulk of evidence was on the side of psi powers being real. During this time Rhine had also founded the *Journal of Parapsychology*, which was where many of his research articles (as well as others supportive of psi phenomena) were first published.

From that point forward, there have been many well-trained, respectable scientists from various fields who have published psi-positive results, using a variety of methodologies (Alcock, 2011). The zenith of such work occurred in the 1960s through 1980s, although it has continued in lesser amounts to the present day (most has moved from academic institutions to private research centers). For example, physicist Helmut Schmidt's work showing that humans and animals could influence random number generators using only their minds and Targ and Puthoff's findings that supported the reality of remote viewing were both highly influential in the 1970s. Robert Jahn, a respected Princeton engineering professor, claimed to have found evidence for psi and published articles on this in the 1980s and 1990s. The most frequent field of study for those examining psi, however, has been psychology. Indeed, the most notable studies to have come out in support of psi's reality in the past two decades were spearheaded by social psychologist Daryl Bem.

Bem, who primarily taught at Cornell University before his retirement in 2007, is one of the most influential social psychologists of the past 50 years, primarily for his self-perception theory of attitude change (which focuses on how behavioral change, even if forced, can result in changes in attitudes). However, in the late 1980s he began engaging in research into psychic phenomena (Bhattacharjee, 2012). His first publication related to psi involved teaming up with Charles Honorton, a prominent parapsychologist. In it, they discussed and defended what's known as Ganzfeld experiments. Their work was published in one of the most widely read and impactful psychology journals, *Psychological Bulletin* (Bem & Honorton, 1994).

In a typical Ganzfeld protocol, the person being tested for psychic powers (the "receiver") is placed in a room alone and a mild amount of sensory deprivation is induced. The idea behind this is that reducing typical sensory input enhances psi powers. This is usually done through the combination of reclining in a very padded chair, listening to white noise (static), and having the eyes covered with halved ping-pong balls while red light is shone into them. Another person (the "sender") outside of that room then attempts to mentally transmit certain information (for instance, what randomly chosen videos he or she is seeing on a television screen) to the receiver for half an hour, who verbally describes what it is that he or she "receives" mentally. Afterward, the receiver listens to this description as recorded by the experimenter and is presented with four potential choices of "target," one of which was supposedly transmitted and three decoys; the receiver must then choose the correct target. This protocol is then repeated several dozen times with a single participant to see what percentage of correct targets are identified and to determine whether the rate is above 25% (the chance rate for guessing).

Bem and Honorton (1994) reported that in their review of the past 20 years worth of Ganzfeld studies they found that there was strong evidence for both the technique being very sound experimentally and that the data from numerous well-controlled studies were persuasive in demonstrating the presence of accurate hits at a higher than expected rate of 32% to 35%. In other words, they were convinced that the data showed clear evidence of paranormal, psychic phenomena—particularly in the case of telepathy. They also found a number of variables related to higher success rates, such as prior belief in psi phenomena, high creativity or artistic skill, and extraversion.

Bem's next major contribution to parapsychology and psi research would come in 2011, when he published an article detailing nine original experiments that he had conducted to test for the presence of precognitive powers. Published in the prestigious *Journal of Personality and Social Psychology*, Bem (2011) sought to test whether stimuli presented after the fact could influence one's current behavior. To do so, he (quite cleverly, it must be admitted) reversed the order of a number of frequently used, well-understood psychological tests. For example:

> In a typical affective priming experiment, participants are asked to judge as quickly as they can whether a picture is pleasant or unpleasant, and their response time is measured. Just before the picture appears, a positive or negative word (e.g., "beautiful," or "ugly") is flashed briefly on the screen; this word is called the prime. Individuals typically respond more quickly when the valences of the prime and the picture are congruent (both are positive or both are negative) than when they are incongruent. In our retroactive version of the procedure, the prime appeared after rather than before participants made their judgments of the pictures. (p. 413)

Of the nine experiments reported on in this article, Bem wrote that all but one showed statistically significant evidence supporting the existence of precognition. That is, he felt that his results showed retrocausality—that future, at the time unknown outcomes, were influencing the behavior of the participants. Needless to say, an article by a prominent and well-respected scientist, published in a top-tier scientific journal, prompted an enormous amount of discussion, including an editorial in the same issue calling for replication efforts (which, as you remember from Chapter 4, are a critical hallmark of good science!).

CRITICALLY EXAMINING PSI POSITIVE RESEARCH

Although this is merely a brief overview of the earliest and most recent efforts into scientifically examining psi of different kinds, you can see that over 80 years of various research projects have had positive findings. Given this breadth and depth of work, one might ask "With all this research, why am I reading about this in a book criticizing pseudoscience?" Well, despite almost nine decades of scientific research into psychic phenomena, most scientists do not consider psi to be established as real. In fact, in the National Academy of Sciences (membership in which is one of the highest honors a U.S. scientist can have), 98% thought that psi had not been scientifically demonstrated (McConnell & Clark, 1991). To understand why almost all mainstream scientists from the physical and social science realms consider the evidence less than compelling requires bringing one's critical-thinking skills to bear on both the individual studies on psi and the proposed theories about how such powers would operate.

Thinking back to our basic critical-thinking tools, several immediately stand out as being highly applicable to claims on psi and psychic powers. Given that there is a strong history of positive research findings, *replicability* and *ruling out rival hypotheses* are the first areas we need to examine. In other words, can these psi-positive results be independently repeated by other researchers? Are there other plausible explanations for why these positive results were found, outside of actual psychic abilities? As we will see, the answer to these questions are "no" and "yes," respectively.

To be accurate, Rhine's early results demonstrating telepathy caused quite a stir in the scientific community, with numerous other researchers attempting to replicate his results. Indeed, scientists from at least five other universities ran the experiments described by Rhine prior to 1940, each time finding no evidence of ESP in thousands upon thousands of trials and subjects (Hansel, 1985). Given this lack of replicability, it becomes critical to find plausible reasons why Rhine was able to find psi when no one else could. As discussed extensively in Chapter 5, our brains are wonderful at ignoring information at odds with what we already believe, and this seems particularly true for Rhine. Indeed, close examination of how the early Rhine experiments were run shows them to be poorly controlled and poorly analyzed, leaving us with a plethora of rival hypotheses that could explain his findings.

As documented in numerous articles and books over the last 80 years, there were a number of methodological problems with these studies. First was that enormous sensory leakage was occurring in the experiments, meaning that the participants had numerous nonpsychic means they could use to figure out what card was being seen by the examiner. Perhaps the most damning was that the cards were printed on such thin paper stock that the symbols on the front could often be seen when looking at the back of the cards. Further, they were in such close quarters that the symbols could also be seen as reflections in the eyes and eyeglasses of the examiners. There is also the very real possibility that the experimenters were, unknowingly, giving subtle cues to the participants about whether they were guessing correctly or not.[2] The cards were also used repeatedly, meaning that they came to have bent corners or other visible damage that could be used to identify particular cards, especially when running hundreds of trials with a single subject. These flaws, so obvious to outsiders, were not seen by Rhine and his colleagues until expressly pointed out. When Rhine's own laboratory put more stringent controls in place to address such issues, they were unable to find any subjects or groups that scored higher than chance levels—findings that were similar to those of other independent researchers.

This pattern of poor experimental controls, methodological flaws, and lack of replicability has surfaced time and time again in parapsychology research. Rhine's studies using dice to examine telekinetic powers, the early Ganzfeld experiments conducted by Honorton, remote-viewing studies funded by the U.S. government in the 1970s, and essentially all of the other well-known psi-positive studies have massive flaws that prevent them from being taken seriously. Furthermore, replication attempts for each fail time and again when conducted by independent parties. Even Bem's work, despite his strong scientific reputation and the high quality of the journals he published them in, shows massive flaws when closely examined.

Experimentally, other researchers have pointed out serious problems with Bem's methods in his precognition experiments, the most problematic being that he (to use a phrase your first author heard numerous times growing up in rural Oklahoma) switched horses midstream. Specifically, in several experiments he appears to have changed his methodology *in the middle of running the experiment* (Alcock, 2011). In other words, he actually

---

[2] This phenomenon was most famously observed in the case of "Clever Hans," a horse that could supposedly count and answer simple math questions by stomping his foot the correct number of times. In the 1890s, Hans and his owner became famous, until it was exposed that Hans was simply responding to subtle physical cues his owner would give when he reached the correct answer. These cues would then prompt Hans to stop stomping. Blinding the owner to the question would result in Hans not giving any correct answers.

conducted two different experiments but lumped their data together for comparison. This was despite the chance probabilities being different for each method. Further, careful reading of the methods reveal that males and females responded differently to different stimuli, so he then allowed them to choose what kind of stimuli they wanted to see, massively biasing the results. And all of these problems show up when taking a careful look at only the first of nine reported experiments! To wit,

> It is difficult to have confidence that the other eight experiments, some of which were carried out earlier than the one just described, were conducted with appropriate attention to experimental rigor: We have toured the laboratory; we have found the dirty test tubes and the mislabeled vials; we have observed inappropriate methodology and analysis. We have lost confidence in the chemist, and there seems little need to poke about further.
>
> Nonetheless, go on we must. (Alcock, 2011)

The rest of the experiments fall prey to many of the same problems as most psi-positive work—suffering from methodological flaws or analytic problems.[3] Further, multiple replication attempts for Bem's work in the years after publication have found no evidence of precognition (Frazier, 2013). As emphasized across the first half of this book, the use of sound scientific methods (including proper controls such as double blinding, sound analytic techniques, and relying on independent verification and replication of results) help prevent us from fooling ourselves into seeing what we want to be there rather than what actually is. The excitement that surrounds new, amazing experimental psi findings has always turned out to be much ado about nothing, once less potentially biased minds examine the experimental protocols and rerun them.

FOOL ME ONCE, SHAME ON YOU; FOOL ME TWICE, SHAME ON ME

Our biases toward evidence that confirms what we already believe can strongly influence how we design and analyze studies, but can also greatly influence how we get information. As such, proper controls can prevent others from fooling us, which has been an all too frequent occurrence for those claiming to have psychic abilities. Dating back to Lucian exposing Alexander almost 2,000 years ago, skeptics have frequently helped others realize when tricksters were taking them in. Master magician and

[3] Grasping the analytic problems with these and other parapsychological studies tends to require a very deep understanding of statistics and as such will not be a focus in this book. However, interested readers are referred to classic works by Gilovich (1991) and Hyman (1989) for an in-depth look at this area.

escape artist Harry Houdini spent a large portion of his later years debunking spiritualists and others who claimed to be able to speak to (and for) the dead (Dunning, 2014). He was, after all, quite familiar with how they accomplished these seemingly amazing feats . . . because early in his career he had done the same show in a traveling circus!

In fact, Houdini went so far as to do a private demonstration of how such feats could be accomplished for his then-friend Sir Arthur Conan Doyle (who, as you remember from earlier in this chapter, inspired J. B. Rhine to give up botany and focus his research on psi). Despite Houdini duplicating the most common feats of mediums of the day, Doyle remained unconvinced that there were prosaic explanations and instead trumpeted supernatural ones. He insisted on having a séance in which Lady Doyle, his wife and a supposed channeler of the dead, would "contact" Houdini's dead mother. Predictably, Lady Doyle's channeling of Houdini's mother failed to impress him, given that the message was both in English (which his mother spoke not a word of) and remarkably generic. As quoted in Polidoro (2001), the first part of the message was:

> *Oh, my darling, my darling, thank God at last I am through. I've tried, oh, so often. Now I am happy. Of course, I want to talk to my boy, my own beloved boy . . . I want him only to know that—that—I have bridged the gulf—that is what I wanted—oh, so much. Now I can rest in peace.*

Houdini, as a good critical thinker, easily applied the principle of parsimony to this situation. Either his mother had actually communicated from beyond . . . or Lady Doyle was (purposefully or via self-deception) making the whole thing up. Accepting the first hypothesis requires us to make numerous nonsupported assumptions—that communication is possible, that his mother learned English in the hereafter, and that all she wanted to say to her grieving child was what anyone could easily guess—leaving it open to being thoroughly sliced through by Occam's razor. The second requires no assumptions at all, given the vast preponderance of evidence showing that people both fool others and themselves regularly by pretending to have psychic powers.

Doyle and Houdini's friendship eventually disintegrated completely after Houdini became active on a *Scientific American* committee designed to investigate such claims of paranormal powers, offering a cash prize of $5,000 (about $70,000 in today's money) to anyone who could demonstrate psi under controlled circumstances. His investigations repeatedly showed that the so-called mediums were using magician's tricks and other deceptions to make it appear that tables were floating, spirits were ringing bells while the medium's hands were still being held, and the like (see Figures 8.2 and 8.3 from Flammarion's 1907 book).

FIGURE 8.2 Table floating.

Source: Reproduced from Flammarion, C. (1907). Retrieved from https://archive.org/details/mysteriouspsychi00flam

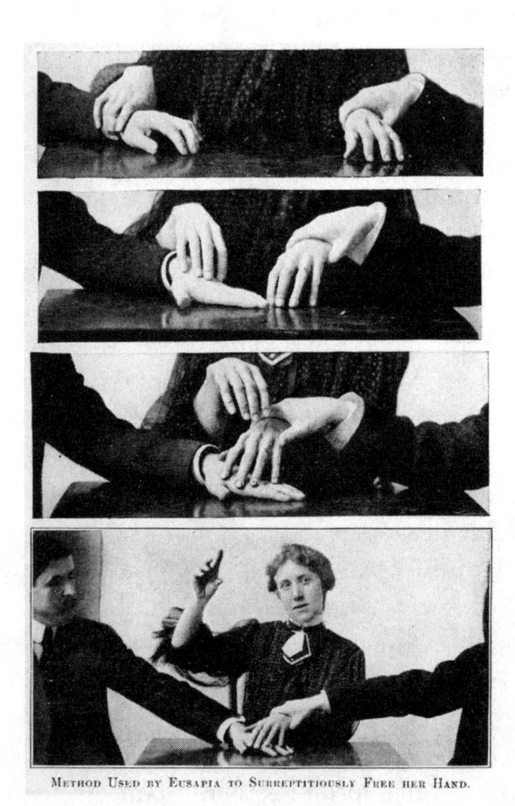

Method Used by Eusapia to Surreptitiously Free her Hand.

FIGURE 8.3 Method used by Eusapia to surreptitiously free her hand.

Source: Reproduced from Flammarion, C. (1907). Retrieved from https://archive.org/details/mysteriouspsychi00flam

After exposing as a fraud someone Doyle had personally endorsed, Houdini was threatened with numerous lawsuits and received threats to his safety. But this did not stop him from both looking for genuine evidence of paranormal powers and finding as fake all of those he investigated. Figure 8.4 is a poster advertising one of the shows in which he demonstrated to people how these "psychics" and "mediums" worked. Houdini was a tireless and valiant skeptic and public educator who did his best to keep people from being taken advantage of until his untimely death, in 1926, at age 52.

In one of the most notable and well-documented cases of researchers being fooled into thinking they had found someone with genuine psychic powers, it was actually the skeptics who did the fooling! Magician James "The Amazing" Randi had begun actively investigating and challenging psychics in the early 1970s after a highly successful career as a stage magician and performer. Much like Houdini before him, Randi offered large sums of

FIGURE 8.4 Houdini poster.

Source: Courtesy of the Library of Congress. Reproduction number LC-DIG-var-1627.

money to anyone who could demonstrate psi under controlled conditions.[4] He is also an expert on the most common conjuring tricks used to make it appear as if one had psychic powers. So when he learned of a large, privately funded grant that was made to Washington University for the express purpose of investigating psychic powers, he set into motion a plan he had thought about for quite some time, dubbed "Project Alpha" (Randi, 1983).

The McDonnell Laboratory for Psychical Research, named after the man who donated the $500,000 to fund it, put out a call for people who thought they had psychic powers to come and get tested under the guidance of lab director and physicist Dr. Peter Phillips. Randi asked two teenage magicians whom he had been exchanging letters with, Steve Shaw and Michael Edwards, to write to Phillips claiming that they had psychic powers. He also wrote to Phillips, sending a list of controls that the experimenters would need to have in place to make sure competent magicians weren't fooling them and even volunteering to observe the experiments at his own expense and for no credit.

Interestingly, although Randi's offer and advice were ignored, the two young men he had coached were chosen as test subjects and put through the experimental protocols for about 4 years and around 160 hours of tests. They duplicated numerous psi abilities using mentalist tricks, learned either by themselves or taught by Randi, such as knowing the contents of sealed envelopes or bending metal objects, fooling the director and other lab personnel into believing they had real psychic powers. Throughout these tests, Randi repeatedly wrote to Phillips, saying that these young men were frauds and suggesting ways to control for tricks. He was summarily ignored until the "results" of the experiment were presented at the PA's yearly conference and Randi showed how these feats were being done, using videos of Shaw and Edwards supplied by Phillips. At that point the McDonnell Lab researchers put proper controls into place to prevent trickery (mostly the ones suggested by Randi some years earlier). Unsurprisingly, the two "psychics" were now unable to perform any of the wondrous feats that had occurred so easily beforehand.

Shortly thereafter, Randi and the young men revealed to the McDonnell researchers that they had been hoaxing them all along. They received praise from both inside and outside the parapsychological community for helping to show the poor experimental rigor that was often applied in such research, but were also criticized both from those who were fooled and by some who felt the deception was unethical. The need for scientists studying psi to collaborate with magicians, mentalists, and others who were trained to purposefully deceive was driven home by Project Alpha . . . but still the

[4] Randi started at $1,000 in 1964, moving to $10,000 by the 1980s. Today the organization he founded—the James Randi Educational Foundation (JREF)—offers a prize of one million dollars to anyone able to demonstrate their powers in controlled, agreed upon conditions (randi.org, n.d.). Despite the thousands of applicants, and several famous celebrity psychics agreeing to take the challenge (but later backing out), no one has come close to winning it.

lesson was not learned by all. One of the primary experimenters from the McDonnell Lab, Walter Uphoff, insisted that Shaw and Edwards actually had psychic powers and just weren't aware of this fact!

Unfortunately this denial of evidence, rationalizing of failures, and lack of critical thinking continues to pervade the parapsychological research community. To take but one example, here is an excerpt from the obituary about one of the McDonnell lab's assistants, Michael Thalbourne, who went on to become well known in that community. It presents a somewhat different point of view on the Project Alpha affair:

> He then took up a post as a junior parapsychologist in the McDonnell Laboratory for Psychical Research at Washington University in St. Louis, MO, working under the supervision of Dr. Peter Phillips. It was in the so-called "Mac-Lab" that Michael's colleagues came under attack by the infamous and now ailing hard-nosed skeptic and magician James Randi, who sought to discredit Mac-Lab staff just to prove some vague point. He had two of his poorly trained magicians infiltrate the lab so they could indulge in deceptive practices that included cheating and all manner of fraud. Needless to say, this reprehensible behaviour was detected, and no published works ever emerged from Randi's farcical escapade (Storm, 2011).

Ironically, though highly critical of the way Project Alpha was conducted, Thalbourne himself (1995) acknowledged that the whole affair contained a major lesson for parapsychologists: that magicians could and should be employed in psi experiments to help prevent purposeful cheating and deception. Yet how often is this actually done? On the side of skeptics and scientific investigators the answer seems to be "increasingly often," whereas on the side of parapsychologists and believers it still appears to be a rare phenomenon. Our willingness to be fooled by both ourselves and others around us when observations confirm and support our already held beliefs appears to have almost no limit. We humans will ignore, argue with, and explain away contradictory information, with this tendency especially true the more personally important or emotionally charged our beliefs are (as discussed in Chapters 5 and 6). In the realm of psi, this is perhaps most apparent when discussing mediumship or ADC.

DEATH, TAXES, AND MEDIUMS USING COLD READING

Dying is the price we pay for being born. Each and every person reading this book will die. Along the way we will lose loved ones—family, friends, mentors, pets—never getting to hold them, speak to them, or in any way interact with them again. In fact, one of the most robust research findings of the past

50 years is that the single most stressful life event you can experience is the death of your child or spouse. Is it any wonder, then, that many people desperately want to speak to those they have lost? Could emotions run any higher and motivations be any stronger than when you are immensely grieving? Speaking personally, I (CWL) would give almost anything to be able to speak to my grandparents, other relatives, and friends who have died, and I'm sure most of you reading feel the same way. If you believe the claims of numerous people today, then I don't just have to wish for this, I can actually make it happen.

From the religious movement inspiring Fox sisters and Madam Helena Blavatsky in the 19th century to the television-viewing inspiring John Edward of *Crossing Over* and Teresa Caputo of *Long Island Medium* in the 21st century, mediums have often acquired huge, devoted audiences by supposedly being able to speak to the dead. This generally translates into financial success, as they often charge people for "readings" or at the very least graciously accept donations. For example, at the time of this writing Edward charges $850 for a private reading, either at his office or over the phone, or $175 to attend a group reading with hundreds of other people (with no guarantee of getting a reading; Edward, n.d.). Caputo has had two best-selling books, sells a line of "spiritually inspired" jewelry, and has a television show that's been airing for seven seasons and averages over 1.5 million viewers per episode.

Given the massive amount of exposure, publicity, and profits made by these and others (James Van Praagh, the late Sylvia Browne, etc.), one would expect that such powers have been thoroughly tested and found to be real. And indeed, they have certainly been tested, with the reality not being kind to the mediums. Even more so than other types of psi, the history of mediumship is rife with exposed frauds and hoaxes (Polidoro, 2003). None of the historical mediums ever investigated were found to have actually been speaking to the dead, and were instead just relying on sleight of hand, tricks, and outright fraud. For example, the Fox sisters, the first major celebrity mediums and the inspiration for the spiritualist movement Doyle would later champion, were famous for the "rapping" sounds that spirits would make in their presence to answer questions . . . as long as you were in a dark room. They finally admitted to having lied and deceived others regarding their "powers" after being exposed by numerous investigators. As many had suspected, the "raps" were produced through a combination of cracking their toe joints and hitting the floor and table with their elbows and knees (Houdini, 1924/2011).

Although today's mediums do not tend to rely on darkened rooms to hide their actions, they do rely on a set of well-documented, well-understood techniques for "reading" people. Typically called cold reading and hot reading, these tools exploit many of the natural biases of the human mind, require no supernatural powers, and can be learned by anyone. To use Rowland's (2002) definition,

> Cold reading is a deceptive psychological strategy. Among other things, it can be used by someone who is not psychic to give what seems to be a very convincing psychic reading. Cold reading is neither one single technique, nor one single procedure. It is better to think of "cold reading" as the collective term for a set of techniques which can be used in different contexts to achieve different goals. (p. 16)

Although cold reading can be used to mimic a number of psi powers (including tarot card or palm readings), we will focus on how it can be used to make it seem that you are communicating with the dead. Back in Chapter 3, you learned about a peculiar aspect of human psychology that's become known as the Forer effect. As a reminder, this is when one is given statements from a trusted source that she believes only apply to her, and are primarily positive; people rate these vague statements (that could apply to anyone) as being very accurate. Exploiting the Forer effect is a key aspect of being a good cold reader, but it is far from the only tool used (Hyman, 1977).

In his guide to learning cold-reading skills, Ian Rowland (2002) outlines seven aspects to making a cold reading work. First is "The Setup," which shows how to set one's self up for giving an excellent cold reading. Actions typically taken by the medium would include being highly personable and likable, encouraging cooperation and volunteering of information, and establishing a ritualistic atmosphere that is intimate but also dissuades challenges. The medium will then attempt to convey that he or she is experienced and knowledgeable and discuss exactly how the process "works." One particularly important part of the setup is making excuses in advance for any failures that may occur.

An example of the "Set Up" can be seen in the television programs hosted by the medium John Edward, including *Crossing Over*. You can see exact examples of John Edward in action on YouTube, but typically in the opening of programs such as *Crossing Over* Edward will explain to his studio audience (and the viewer) what he will do and say during the show. He will make use of specific terminology regarding the information that he "receives." For example, he will refer to people who are older than the individual receiving a reading as being "above" the individual. He is also careful to say that when he receives the names of persons who have "crossed over," he may be "a little bit off" on the way a name sounds, but assures the audience that he will be correct with part of the name, such as the first initial. He will usually advise his audience that their role is to confirm for him the information he is receiving from those who have "crossed over" and to "let what happens happen."

The confidence and surety the medium exudes during this type of opening spiel also helps to set the tone of the reading, as research has shown that the more confident you are the more likely you are to be believed (Hyman, 1977). But, you also want to balance this with a sense of humility while still asserting your confidence.

After settling people in, mediums will then move into the main portion of the reading. Although the techniques can differ slightly depending on if they are being performed in a one-on-one or group setting, it generally begins with a barrage of statements and/or questions thrown out—the so-called "shotgun" move. You are hoping that by saying enough vague things that one of them will be picked up by someone in the audience. When that one is picked up, you then immediately drop everything else that was said and focus your attention on that. So, the cold reader supplies the words, but the person being read applies meaning to those words. For example, in a typical session from *Crossing Over* John Edward might say that he feels a "younger female energy" coming through within a group of people in his studio audience, stating that the energy identifies itself as a daughter, niece, or granddaughter. He might refer to a "connection" with a specific physical ailment such as cancer, and ask the group for verification. He might then proceed to throw out some similar-sounding female first names that he is "getting" such as Margaret, Maggie, Marge, even saying that the name simply starts with an "M."

Someone will, undoubtedly, "bite" for these kinds of broad statements. Maybe a woman raises her hand, revealing that her deceased mother's name was Margaret. Edward would then pounce on that, plunging ahead with more broad statements, saying things like "she wasn't doing what she would normally do" on the day she died. Making statements such as these can be called "fuzzy facts" because they are very likely to be accepted and you can then spin them into specifics. For example, the medium might say "I'm getting that there was something to do with the chest," which can then be spun into a heart attack, or congestive heart failure, or lung and breathing difficulties, or numerous other things, depending on how someone responds. But where was the "younger female energy" he was so confident about? Or the cancer? Both ignored and forgotten, just another in a series of misses left by the wayside.

You can also use statements that can be interpreted as being highly specific, but in fact just rely on statistical or trivia-type knowledge about people. For example, the cold reader could say "He's telling me something about jewelry, about broken jewelry" or "She's saying something about the dress she never wore." Almost all households will have some piece of broken jewelry in them, and almost all people have bought some clothing that they never actually wore. But for someone really looking for communication from the dead, these broad statements can be latched onto and seen as "evidence" of their loved ones talking through the medium. The use of these vague, broad statements that rely on common aspects of people's lives is so common and repetitive among mediums that one person even developed a "Cold Reader Bingo" card (see Figure 8.5) to use when watching these performances, so you can keep track of them (Skeptico, 2007).

Although there are numerous ways of asking questions, using observation of physical characteristics, and particular themes that are usually elaborated on (love, money, concern about the future, unresolved conflicts with those that have died), this is the crux of a good cold reading: set the stage appropriately, instill confidence, make pre-excuses for any misses, and use lots of questions. The skilled cold reader is also good at turning misses into hits via cleverly changing the content of what was said or by even blaming the person being read for not understanding. This can often be worked into the opening spiel, the stage setting. For example, in a taped interview (*Penn & Teller: Bullshit*, 2003) John Edward once stated that he is not accurate 100% of the time, but what audience members may initially think are "misses" in a reading are not. He always "gets it" during the reading and tries his best to interpret "it" for the person on the other end of the reading.

Finally, remember what Hyman (1977) calls "the golden rule" of cold reading—always tell the client what he wants to hear. For mediums, this almost invariably ends up being some variation of "They love/miss/forgive you, and want you to move on with your life and be happy." You rarely see a medium giving bad news to a grieving family member ("Wow, I am sorry to report this, but he is currently burning in the flaming fires of Hell"),

COLD READER BINGO				
"J" (or "J sounding"—includes "G") name	"Yes you do!"	Father figure/ older male	Show host praises guest—"wow!" or similar	"Chest area" or "Breathing trouble"
"Birthday" "Wedding"	"Boxes"	"Cancer"	Wild-ass guess	Any number from 1 to 12
"S" name	"Write this down"	"R" name (includes "Bob")	"Jewelry"	Asks about child or "Toys"
Person accepts miss as a hit	Multiple fishing questions	Complete miss on all guesses for a person	Dog or cat	Mother figure/ older female
"Head area"	"M" name	"Badge" "Flag"	The dead relative is "OK" "Fine"	"Do you understand?"

FIGURE 8.5 Cold Reader Bingo.

Source: Skeptico, 2007 (Weblog). Reproduced with permission.

especially if there's a chance that person may return for another reading (and hence another paycheck for the medium).

If you watch several demonstrations of so-called mediumship (and examples abound, just type the name of any of those mentioned in this section into YouTube), you can see these cold-reading techniques demonstrated time and time again. General information playing on the Forer effect, confirmation biases fully displayed by forgetting the misses and the information that was fed to rather than received from the medium, and people's high motivation to connect with those who have died all mash together for a very convincing demonstration of psi, as long as you don't have your critical-thinking skills in place.

But what about extremely detailed and accurate information given to an individual during a reading? How could that possibly be explained outside of actual psychic powers? In many cases this is the result of what we call "hot readings." Hot readings occur when the medium has gained information beforehand and then uses it during their reading, acting as if they obtained it from a dead person. There are numerous ways this can be accomplished, with one of the most common involving an accomplice who circulates in the audience before the reading, talking to people about who they are wanting to contact, how that person may have died, or listening in on conversations and then reporting back to the medium before the show begins.

For instance, in a Dateline NBC investigation of John Edward, they found that he had spoken to a cameraman extensively before the scheduled shoot and then later used that information and claimed it was the result of an "otherworldly revelation" (Nickell, 2010a). An earlier *TIME* article had found that Edward had numerous assistants milling around before the show, talking to people, having them fill out name cards that included family history information, and then helping them to preassigned seats (Jaroff, 2001). It makes one wonder why, if he was truly getting messages from beyond the grave, any of that would need to take place. To quote Penn Jillette from the first episode of the Showtime Series *Penn & Teller: Bullshit!* (2003):

> Why are these psychics ever wrong? If you were ever really talking to the dead, why would they ever get it even slightly wrong? Are the dead watching TV while they are communicating and so distracted that they can't follow the conversation? Do they mumble when they get to the details?

CONCLUSIONS

From as far back as I (CWL) can remember, I have been a huge fan of superheroes, science fiction, and fantasy. From the telepathy of the X-Men's Professor X to Stephen King's nightmarish telekinetic in *Carrie* to the psychic powers of the Jedi and Sith in *Star Wars*, characters able to do amazing feats with only the "power of their minds" hold a special place in my heart. However, as much as I personally think it would be amazing to have people who are actually able to communicate with the dead, read minds, or predict the future, the evidence just is not on their side. Instead, we see a consistent pattern of claimed powers that are found to be hoaxes, of psi-positive research not being well controlled or replicable, and of a distinct lack of psychic powers being revealed in the lab. Using scientific methods and critical-thinking skills can lead us to only one answer—psychic powers are indeed all in one's head, but not in the way believers claim.

QUESTIONS FOR REFLECTION

1. *Have you ever had a premonition that something was about to happen, and then it did? How would you explain this phenomenon (aside from believing that you have special powers)?*
2. *If you disbelieve that we can communicate with the dead, what would convince you that we could? Or vice versa, if you think we can communicate with the dead, what would persuade you that we can't?*
3. *Do you consider mediumship harmless entertainment? If not, who do you think should bear the blame for exploitation—the medium or the victim?*
4. *What's your best explanation for why people believe we can communicate with the dead?*
5. *Given the arguments in this chapter, do you think it epistemically responsible to "keep an open mind" about psi, or do you think we can consider the belief that it's false fully justified?*

REFERENCES

Alcock, J. (2011, January 6). Back from the future: Parapsychology and the Bem affair. *Skeptical Inquirer*. Retrieved from http://www.csicop.org/specialarticles/show/back_from_the_future

Bem, D. J. (2011). Feeling the future: Experimental evidence for anomalous retroactive influences on cognition and affect. *Journal of Personality and Social Psychology, 100*(3), 407–425.

Bem, D. J., & Honorton, C. (1994). Does psi exist? Replicable evidence for an anomalous process of information transfer. *Psychological Bulletin, 115*(1), 4–18.

Bhattacharjee, Y. (2012, March). Paranormal circumstances: One influential scientist's quixotic mission to prove ESP exists. *Discover*. Retrieved from http://discovermagazine.com/2012/mar/09-paranormal-circumstances-scientist-mission-esp

Dunning, B. (Producer). (2014, September 2). *Harry Houdini and Sir Arthur Conan Doyle* [Audio podcast]. Skeptoid Media. Retrieved from http://skeptoid.com/episodes/4430

Edgar Cayce's Association for Research and Enlightenment. (n.d.). *Dead sea scrolls*. Retrieved from http://www.edgarcayce.org/are/ancient_mysteries.aspx?id=2667

Edward, J. (n.d.). *John Edward private readings*. Retrieved from http://johnedward.net/events/private-reading/explained

Flammarion, C. (1907). *Mysterious psychic forces: An account of the author's investigations in psychical research, together with those of other European savants*. Boston, MA: Small, Maynard, and Company.

Frazier, K. (2013). Failure to replicate results of Bem parapsychology experiments published by the same journal. *Skeptical Inquirer, 37*(2). Retrieved from http://www.csicop.org/si/show/failure_to_replicate_results_of_bem_parapsychology_experiments_published_by

Gauld, A. (1968). *The founders of psychical research*. London, UK: Routledge & Kegan Paul.

Gilovich, T. (1991). *How we know what isn't so: The fallibility of human reason in everyday life*. New York, NY: The Free Press.

Hansel, C. E. M. (1985). The search for a demonstration of ESP. In P. Kurtz (Ed.), *A skeptic's handbook of parapsychology* (pp. 105–127). Amherst, NY: Prometheus Books.

Houdini, H. (2011). *A magician among the spirits*. New York, NY: Cambridge University Press. (Original work published 1924)

Hyman, R. (1977). Cold reading: How to convince strangers that you know all about them. *Skeptical Inquirer*, 2(1), 18–37.

Hyman, R. (1989). *The elusive quarry: A scientific appraisal of psychical research*. Amherst, NY: Prometheus Books.

Jaroff, L. (2001, March 5). Talking to the dead. *Time, 157*(9), 52.

Lemesurier, R. (2010). *Nostradamus, bibliomancer: The man, the myth, the truth*. Wayne, NJ: New Page Books.

Loxton, D. (2012). Dark secrets of the oracle-monger. *Skeptic, 17*(14), 65–72.

McConnell, R. A., & Clark, T. K. (1991). National academy of sciences' opinion on parapsychology. *Journal of the American Society for Psychical Research, 85*, 333–365.

Nickell, J. (2010a). John Edward: Spirit huckster. *Skeptical Inquirer, 34*(2). Retrieved from http://www.csicop.org/si/show/john_edward_spirit_huckster

Nickell, J. (2010b). Nostradamus: A new look at an old seer. *Skeptical Inquirer, 34*(5). Retrieved from http://www.csicop.org/si/show/nostradamus_a_new_look_at_an_old_seer

O'Keeffe, C., & Wiseman, R. (2005). Testing alleged mediumship: Methods and results. *British Journal of Psychology, 96*(2), 165–179.

Penn & Teller: Bullshit! (2003, January 24). Season 1, Episode 1 [Television series]. Director, S. Price. New York, NY: Showtime Networks, CBS Entertainment. Retrieved from http://www.tv.com/shows/penn-teller-bullshit/talking-to-the-dead-238519

Polidoro, M. (2001). *Final séance: The strange friendship between Houdini and Conan Doyle*. Amherst, NY: Prometheus Books.

Polidoro, M. (2003). *Secrets of the psychics: Investigating paranormal claims*. Amherst, NY: Prometheus Books.

Randi, J. (1983). The project alpha experiment: Part 1. The first two years. *Skeptical Inquirer, 7*(2). Retrieved from https://archive.org/stream/JamesRandiTheProjectAlpha/JamesRandi-TheProjectAlphaExperiment-SkepticalInquirer1983_djvu.txt

Randi.org. (n.d.). *The million dollar challenge*. Retrieved from http://web.randi.org/the-million-dollar-challenge.html

Rhine, J. B. (1927). One evening's observation on the Margery mediumship. *Journal of Abnormal and Social Psychology, 21*, 401–421.

Rhine, J. B. (1934). Extra-sensory perception of the clairvoyant type. *Journal of Abnormal and Social Psychology, 29*, 151–171.

Rhine, J. B., Pratt, G., & Stuart, C. (1940). *Extra-sensory perception after sixty years: A critical appraisal of the research in extra sensory perception*. New York, NY: Holt.

Ronson, J. (2004). *The men who stare at goats*. New York, NY: Simon & Schuster.

Rowland, I. (2002). *Full facts of cold reading*. Surrey, UK: Ian Rowland Limited.

Shermer, M. (2011). *The believing brain: From ghosts to gods to politics and conspiracies—How we construct beliefs and reinforce them as truths*. New York, NY: St. Martin's Griffin.

Skeptico. (2007). *John Edward/James van Praagh Bingo*. Retrieved from http://skeptico.blogs.com/skeptico/2007/11/john-edward-jam.html

Storm, L. (2011). Obituary for Michael Thalbourne. *Mindfield, 2*(2). Retrieved from http://www.parapsych.org/articles/29/123/2011_outstanding_contribution.aspx

Thalbourne, M. A. (1995). Science versus showmanship: A history of the Randi hoax. *Journal of the American Society for Psychical Research, 89*, 344–366.

Varvoglis, M. (n.d.). *What is psi? What isn't?* Retrieved from http://archived.parapsych.org/what_is_psi_varvoglis.htm

CHAPTER 9

UNKNOWN ANIMALS AND CRYPTOZOOLOGY

In art and stories dating back to the dawn of our civilization, monsters of one kind or another have played a significant role. From European cave paintings of fantastical creatures dating back 40,000 years to the Sumerian *Epic of Gilgamesh*'s Humbaba of 2,100 BCE, the hydra and minotaur of Greek mythology, and the thunderbird and wendigo of North America's native peoples, monsters have long fascinated us and provided both fodder for epic tales and frights for campfire stories. Although few today truly believe in the existence of an enormous bird that can cause storms or a supernatural beast-man that can possess others and turn them into cannibals, there are still many tales of monsters floating around under a new, more scientific-sounding name: cryptids.

Cryptids are the focus of study in cryptozoology, a field most scientists label as pseudoscientific. Referring to the study of "hidden animals," cryptozoology is sometimes called a "discovery science" by proponents, in that it is not focused on research in a traditional sense but instead on exploration and documentation (Arment, 2004). Focusing on animals whose current existence is not supported by evidence, whether those are extinct or legendary animals, cryptozoologists purport to try to demystify these mystery creatures using rigorous examination and evaluation. However, when looked at through a critical-thinking lens, cryptozoology is more an extension of our species' long obsession with the dangerous and unknown, updated for the modern era, and less a scientific endeavor.

Cryptozoology is a relatively modern invention since, prior to the 1900s, the discovery of large creatures undocumented by science was a relatively frequent event. Particularly as European colonists and researchers poured into Africa and the Americas, documentation of "new" animals (which were, of course, well known to the indigenous populations) were almost a common occurrence for explorers. But, as maps of the world became increasingly filled in and potential habitats for new animals shrank, the discovery of megafauna (large vertebrates) drastically slowed and professional scientists replaced the eager amateurs of the past (Dendle, 2006). This isn't to say that no new animals have been found, of course. Land animals, such as the okapi and mountain gorilla of Central Africa, only had their existence confirmed in the early 1900s after decades of rumors. The over 5-foot-long coelacanth was rediscovered in the waters off South Africa in 1938 after having been thought to be extinct since the end of the Cretaceous period some 66 million years ago. The colossal squids, long a staple of frightening stories told by sailors, have only recently been well documented, despite sizes up to 4.5 meters (15 feet) long and weights of over 500 kilograms (1,100 pounds)!

Given these fantastic discoveries, is it so hard to believe that other large, previously undocumented animals await intrepid adventurers and explorers? This is the contention of cryptozoologists, who tend to focus on these animals rather than searching for relatively unexciting things like new species of beetles. In this chapter, we take a look at some of the more popular cryptids, the support for and against their existence, and compare the careful methods and types of evidence used by scientists to those most often relied on by cryptozoologists. Because of length constraints, this chapter can describe only a few of the more popular and well-known cryptids from the hundreds of possible suspects (see Coleman & Clark [1999] for an exhaustive overview). We look at some of the more well-known cryptids around the world: the Sasquatch or Bigfoot of North America, the Loch Ness Monster of Scotland, the Mokele-Mbembe of Africa, and the Chupacabra of Mexico and South America.

MAN-APES IN NORTH AMERICA

Probably the most well-known cryptid in the world is some type of man-ape or "wild man." Stories exist about such creatures across many cultures using many names, from Nepal's Yeti or Abominable Snowman to Australia's Yowie to China's Yeren to Russia's Chuchunaa. In the Western world such tales date back to ancient times, from the stories of Pliny the Elder 2,000 years ago about small beast-men called the Nittaewo to the woodwose of medieval Europe. In the United States, such creatures tend to be given the name Bigfoot or Sasquatch, although there are regional variations such as "Skunk Ape" in the southeastern United States.[1] Although stories of a large ape-like hominid seem to have begun in the Pacific Northwest among indigenous populations, today people claim to have seen a Bigfoot in every one of the United States except Hawaii and in all eight of the lower Canadian provinces (Bigfoot Field Research Organization, 2015).

In modern lore, the Bigfoot is typically described as seen in Figure 9.1: a large and hulking creature covered in dark hair, usually brown or red-brown, with heights of 2 to 3 meters (6–10 feet) and weights of over 230 kilograms (500 pounds). As befits a primarily nocturnal creature, it reportedly has large eyes under a low-set forehead, and a rounded head.

[1] For ease of communication, the term "Bigfoot" will be used in this chapter to refer to all of these variations on a large, unknown hominid in North America.

FIGURE 9.1 Bigfoot.
Illustration by Ryan Long.

As one might expect given the common name for the cryptid, footprints as large as 60 centimeters long and 20 centimeters wide (2 feet by 8 inches) have been attributed to it (Buhs, 2009). Typically seen alone, there have been some reported sightings of small groups of the creatures as well. The diet is typically thought to be omnivorous, consisting of a variety of fruits, nuts, tubers, and caught prey.

Brief History of Bigfoot Encounters

Although published stories of such creatures can be traced as far back as the mid-1800s, it was really the newspaper articles by J. W. Burns that helped gain publicity for the stories of large hominids living in the Pacific Northwest (Loxton & Prothero, 2013). He told stories collected from native people about a race of "hairy giants" that he called Sasquatch, an Anglicized version of a native term for the creatures. The natives and others living in British Columbia (BC) had reportedly had several run-ins with the creatures, who tried to avoid humans. Reports of encounters from that point forward were relatively sparse until the 1950s, when two things happened.

The first event that reintroduced the Sasquatch (the term "Bigfoot" hadn't been coined yet) to public consciousness was a clever marketing campaign by the small town of Harrison Hot Springs in BC. They sponsored a "Sasquatch Hunt" to help celebrate BCs centennial in 1957. The hunt was subsequently picked up by and splashed all over the front page of newspapers not only in Canada, but across the globe. Teams of hunters (and hoaxers) descended on the small town hoping to cash in on the $5,000 prize being offered to anyone bringing one in alive. The attention generated by this publicity stunt was enormous and critical to the second event.

In that same year (and likely as a result of all the publicity) a man named William Roe came forward with details of a meeting with a Sasquatch. He described being in the woods and happening upon a large, hairy, bipedal creature covered head to foot in dark brown hair. After Roe had observed it for a short time, the creature moved away when it noticed him. Apparently, he had encountered a female, as he noticed it had breasts when it turned toward him, and further described long, simian-type arms, a broad and flat nose, and a very thick neck. Roe claimed that this had happened in 1955, but he only came forward after the Harrison Hot Springs hunt became big news. Analogous to the way Betty and Barney Hill's alien encounter sparked the modern abductee story covered in Chapter 7, Roe's sighting and description would become the template for future alleged encounters with the Bigfoot.

At this point, the sightings began to spread beyond the confines of Canada, with Sasquatch reportedly being encountered in Washington State. In 1958, the encounter that would give this mystery creature a new name took place in California. On a work site near Bluff Creek, giant footprints repeatedly appeared overnight after the crew had gone home. Initially taken to be a hoax of some kind, one of the workers took a plaster cast of the big footprints to a local newspaper, and the mystery of "Bigfoot" (as the newspaper called the creature that had supposedly made the tracks) was born. The contractor for the site, Ray Wallace, and several of his employees continued to find new prints and two of them even claimed to have seen a 10-foot- tall creature in the woods that closely resembled the Sasquatch of Roe's story the year before.

From this point, the story of the mysterious man-ape spread like wildfire across the United States and Canada, in large part because of extensive national media coverage of the footprints from Bluff Creek. There were numerous "hunts" for Bigfoot sponsored by various companies and individuals, sightings reported from coast to coast, and plaster casts of giant feet made by the thousands. Things really got a boost in 1967, when the most argued over and debated piece of evidence for the existence of Bigfoot was captured . . . on film!

Two friends, Roger Patterson and Bob Gimlin, had set out to record footage for a documentary about Bigfoot. Given the history of sightings in the area they chose the Bluff Creek region to shoot in, close to where the footprints that ignited America's interest in the creature were first found. Patterson, in particular, had an interest in the creature, having traveled to the region before, interviewing people, and looking for evidence of Bigfoot. He had even self-published a (not particularly original) book about Bigfoot in 1966. In October 1967, he and Gimlin took a rented 16-mm film camera with them into Bluff Creek, where they soon shot one of the most famous short films in history. The film clocks in at under 1 minute, but appears to show a large, humanoid figure covered in dark fur walking slowly away from the camera and turning its head back once to look at the camera. The film is very shaky at first, reportedly because of Patterson was running toward the creature in an attempt to record it, but it stabilizes about halfway through.

To everyone who had read Roe's account of meeting a Bigfoot, this film was extremely familiar. The long arms, tall stature, and short, dark brown fur are all just as he described for the creature he encountered. Moreover, Patterson and Gimlin also took plaster casts of the footprints it had left, showing the expected large feet just as in Ray Wallace's encounters. It seemed that, if not the holy grail of a live body, at least a very revered chalice of Bigfoot had been found.

In the almost 50 years since the Patterson–Gimlin film, large numbers of other plaster casts, footprints, photographs, and videos from supposed encounters with a Bigfoot have surfaced. Some websites, like the Bigfoot Field Researchers Organization and Bigfoot Encounters, list the number of reported sightings over the past 60 years as being over 5,000 in the United States alone, with hundreds occurring each year. Numerous hairs have been found and labeled as being from a Bigfoot, and a journal even printed an article outlining an alleged analysis of Bigfoot's DNA based on hair, blood, and other biological samples. The cable TV network Animal Planet has a show called *Finding Bigfoot* that has run for over six seasons. Given all of this gathered evidence, why are we covering Bigfoot in a chapter on pseudoscience rather than in a zoology textbook?

Critical Thinking Applied to Bigfoot Evidence

Several pieces of the standard Bigfoot story and the evidence for its existence need critical examination. First is how the stories of Bigfoot have evolved over time. Similar to how the descriptions of aliens varied greatly before coalescing over time with increasing media coverage of alleged abductions (as discussed in Chapter 7), we can see that descriptions of Bigfoot differed considerably in the years prior to the mid-1900s. For example, Native American and First Peoples' legends of creatures that have bodies covered in stone, are supernatural cannibals, or are underwater dwarves are sometimes all lumped together as "evidence" of a long oral tradition of the existence of Bigfoot (Loxton & Prothero, 2013). In the original Sasquatch articles written by Burns, the creatures were not covered in hair but instead just *had very long hair on their heads*. In fact, reading those articles gives a description of Sasquatch as just larger versions of human beings who spoke an understandable language, had villages, and wore clothes. But as media coverage about alleged sightings increased starting in the 1950s, the descriptions given by witnesses become more and more uniform, particularly following the man-ape stories of William Roe, the encounters of Ray Wallace, and especially the Patterson–Gimlin film. So you can conclude that all the Bigfoot (or is it Bigfeet?) started looking the same only after 1955, or you can conclude that people's perceptions are colored and influenced by information they have been exposed to previously.

The second major problem is the consistent history of hoaxers in the Bigfoot community. Remember Ray Wallace, the contractor whose employees made the plaster casts that gave rise to the term "Bigfoot" and claimed to have seen the creature in the woods of California? He was a well-known prankster, and 2 years after the plaster casts were made Wallace reported that he had actually captured a Bigfoot, which ate nothing but Frosted Flakes cereal (he was never able to produce the creature, of course). After his death in 2002, Wallace's family came forward with a pair of carved wooden feet, which they said had been used by their father to fake Bigfoot prints across the years. They further stated that he had multiple sets, all a bit different, that had been used in various locations at various times. He was even reported to have collected and left behind hair and feces near some of the faked tracks, in order to further the "evidence" of a Bigfoot population.

This is far from the only time that some Bigfoot prints were revealed as a hoax or someone has claimed to have captured or killed a Bigfoot only to later recant. In fact, in the past decade alone there have been three highly publicized cases of individuals possessing either a live or dead creature. Each time they received major news coverage from all types of press outlets, and each time they were found to be lying. That old saying seems to apply in this case: Fool me once, shame on you; fool me twice, shame on me; fool me for 60 years and I might just believe in Bigfoot.

Many Bigfoot believers will downplay the hoaxes, saying that they are not as numerous as claimed or that they do not detract from all the "real evidence" out there. This lies at the heart of the Bigfoot problem—that there is a distinct lack of *good* evidence for its existence. There are many eyewitness stories, but think about all the known and documented problems with human memory and perception described in Chapter 7. Just as we are bad at identifying objects in the sky at night, we are equally poor at identifying objects in the woods or at a great distance. Famous "sightings" of Bigfoot have even turned out to be tree stumps seen from a particular angle, or (more commonly) bears. In fact, the habitat of black bears in North America overlaps remarkably with the places reporting the most

Bigfoot sightings (Lozier, Aniello, & Hickerson, 2009). Bears are large, hairy creatures that can often be seen moving around on two legs and tend to live in wooded areas where there are few humans. Any of that sound familiar? There is also the problem that in the Bigfoot literature, reported sightings diverge widely and have included pointy Vulcanesque ears, patterns on the fur, and heights ranging from human size to over 12 feet tall across the past 90 years.

The videos and photos of creatures purported to be Bigfoot tend to be so blurry and noninterpretable that the term "blobsquatch" was invented to describe the labeling of nonidentifiable objects such as the cryptid. Even the Patterson–Gimlin film, which many cryptozoologists believe is definitive proof that the creature exists, is very inconclusive. In fact, it's so questionable that skeptics and Bigfoot enthusiasts can't agree on much of anything about it . . . even when talking about it with others from the same background! As Loxton and Prothero (2013) say during their exhaustive analysis of it, "no one knows whether the film depicts a real Sasquatch or a man in a gorilla suit" (p. 44).

There is also a plethora of poor quality physical evidence, but nothing solid. On analysis, all of the hair samples that have been collected have turned out to be synthetic or from known animals or humans. In a recent peer-reviewed study, 30 hair samples that were supposedly from "anomalous primates" were gathered by a lab at Oxford University and analyzed using mitochondrial RNA sequencing (Sykes, Mullis, Hagenmuller, Melton, & Sartori, 2014). All but two of the hairs were from known living animals (bears mostly, but they also included deer, a tapir, and a human) with the remaining two samples coming from some type of unknown bear species that apparently lives in the Himalayas. Although this discovery is in and of itself quite fascinating, it is not what the cryptozoologists had hoped for. Still, it was an excellent example of doing well-designed, ethical, replicable science. Sykes et al. (2014) were very open about how they collected their samples (by asking for leading cryptid hunters from across the globe to send them hairs for analysis) as well as the exact methods they used to analyze the hairs. They went through the peer-review process first and then announced their results to the media after their paper was published. Finally, they disclosed the funding that assisted in paying for the project (Icon Films, who produced the British TV series *The Bigfoot Files*) and published their article in an open-access format so that anyone with access to the Internet could read it. This entire process stands in sharp contrast to another genetic study conducted by a Bigfoot proponent named Melba Ketchum.

Ketchum, a veterinarian who also owns a genetics lab, was at the center of a 5-year-long saga over her analysis of alleged Bigfoot DNA (Hill, 2013). Starting in 2008, she made a number of claims about having done extensive genetic analysis of hair, blood, flesh, and even a toenail that reportedly belonged to a number of Bigfoot. In what amounts to a massive scientific faux pas, she released a press statement in November 2012, prior to the publication of her article, claiming that her team of researchers had definitive evidence of a "novel hominid hybrid species, commonly called 'Bigfoot' or 'Sasquatch,' living in North America" (Hill, 2013). Ketchum also reported that she had a personal encounter with a group of Bigfoot whom she described as playful and gentle and that she knew where they lived. The media had a field day with her release and claims, covering the story extensively despite the lack of a published paper or any available data.

In February 2013, the paper was finally published. However, it was published online in the first issue of a brand-new journal and was the only paper published in that issue. A quick bit of detective work showed that the journal had been registered under Ketchum's name the prior month and that the website had been online for less than 2 weeks before the Bigfoot DNA paper was published. Although the journal was advertised as "open access," in order to access the (sole) article on the site you had to pay $30.[2] Further fishiness abounds as she has refused to grant other researchers access to her samples in order to run replicative analyses (remember how important that is to our knowledge base?).

When actual geneticists got their hands on the paper and its raw data, the general consensus was that they were, to quote one report, "a mess" (Timmer, 2013). Not only was the interpretation of the analyses highly suspect, but the findings contradicted everything that is known about how reproduction and biology (specifically interspecies breeding) works. The techniques were sound, but Ketchum and her colleagues seem to have forgotten one of the most basic aspects of science and critical thinking—falsification. The data, given how huge a set they were working with and how many techniques they had for manipulation at their disposal, could basically be worked into supporting any preformed hypothesis. What they failed to do was set up their methodology to try and disprove their belief that the DNA came from Bigfoot. Instead, they started with a foregone conclusion ("These samples are from Bigfoot") and worked very diligently to make their data fit that conclusion.

So if some of the best "evidence" is the result of hoaxes, the eyewitness reports are not reliable for numerous reasons, and the (very limited) physical samples in no way point toward a new species of hominid, what are we left with? From a scientific viewpoint, the answer is "not much at all." Unless someone comes up with a body (which, given the necessary breeding population to account for the number of sightings, should have occurred in the past 60 years of searching) or some new stunning evidence, critical thinkers have to take a look at the whole shebang and declare that Bigfoot isn't just dead, he was never alive in the first place.

---

[2] Adding to the shadiness of this whole enterprise is that, as of the writing of this chapter (June 2015), only one other article had been "published" in the journal's 2½-year existence. The only author (in a shocking development) was Dr. Melba Ketchum.

LIVING "DINOSAURS" IN AFRICA AND SCOTLAND

Approximately 66 million years ago, Earth experienced a mass extinction event resulting from climate change that was most likely triggered by a massive asteroid impact. Called the Cretaceous–Paleogene extinction event, scientists estimate that 75% of animal and plant species alive at that time were wiped out, including the nonavian dinosaurs and their water-dwelling plesiosaur cousins.[3] But, what if these creatures did *not* all die but have continued living into the modern day, hidden away from society in inaccessible regions of the planet? That is the primary claim underlying the next two major cryptids we are going to cover—the Loch Ness Monster of Scotland and the Mokele-Mbembe of the African Congo.

The Loch Ness Monster

The Loch Ness Monster (often referred to as "Nessie") is a widely known cryptid, second only in popularity to Bigfoot among the general public. Reportedly living in Loch Ness, a huge lake in the Scottish Highlands and the largest freshwater body in the United Kingdom, modern sightings and reports of the creature date back to the early 1930s.[4] By most accounts, Nessie resembles a long-necked plesiosaur, having a small head with a long, narrow, flexible neck attached to a broad, flat body with four powerful flippers and a short tail. The earliest reports of the creature place the size of the body at 7 to 8 meters (25–30 feet) long with an additional 3- to 4-meter (10–12 foot) long neck. Over the past 80 years, there have been several intriguing reports of encounters with the creature, with the first happening in 1933 when a couple driving their car near the lake came upon"a most extraordinary form of animal" (as he wrote in a letter to the local newspaper) crossing the road only 20 yards or so in front of them making its way to the lake (Spicer, 1933).

Several more locals reported seeing it, and over the next decade there were several highly intriguing photographs and videos taken that supposedly showed the creature. Probably the most famous of these was taken in 1934 and called the "Surgeon's photograph." It shows what appears to be the long neck and head of a creature coming out of the lake, looking very similar to the earlier descriptions of a plesiosaur-like creature. More images (using underwater cameras, video cameras, sonar, and submarines) combined with

---

[3] Many people assume that because they were huge animals that went extinct at the same time, that plesiosaurs were a type of dinosaur. They were not. Instead, plesiosaurs were actually marine reptiles that lived contemporaneously with the dinosaurs and died off at the same time.

[4] Some people claim that a story from the 7th century text *Life of Saint Columba* (Adamnan, 1874) records an early encounter with the creature, but that takes place at the River Ness rather than the Loch Ness. Much like most medieval tales of saint's lives, it is also filled with many other fantastic stories so its accuracy is highly doubtful. Some people have even attempted to draw connections between the various water creatures found in native Scottish folklore, such as water-horses and kelpies, and Nessie, but in the stories these creatures are obviously mythical and magical in nature.

FIGURE 9.2 Nessie.
Illustration by Ryan Long.

some shrewd public relations moves to increase the visibility of Nessie and turn it into a worldwide phenomenon that generates millions of dollars in tourist revenue for the region per year (see Figure 9.2). But is it really a prehistoric marine reptile that has survived the millennia or something else entirely? And could there be other relics, like the rediscovered coelacanth, waiting to be found in other remote areas?

Mokele-Mbembe

Moving far to the south and east, we find another story of a large, prehistoric creature, this time living in the swamps and jungles in the heart of Africa. This is the Mokele-Mbembe, an alleged dinosaur (specifically a sauropod like the apatosaurus) that survived the mass extinction of the dinosaurs and has been living in relative secrecy since then. Although scores of monsters exist in the mythology of the central African tribes, it wasn't until the early 20th century that European explorers began sending back tales of an unknown but enormous beast that dwelled in the lakes and rivers and was reportedly well known by the local populace for being quite fearsome. Intriguingly, these tales were not initially confined to a single geographic location but were instead spread across the continent, from Niger in the north to the central Congo Basin to Rhodesia and South Africa at the southern tip of the continent. Similar to the stories of Bigfoot, there were numerous descriptions of the creature(s) that emerged around this time, some clearly inspired by the sauropod fossils that were the rage with the intelligentsia in the early 1900s and others that seemed a mishmash of known creatures like rhinos, pythons, and equines.

From the 1960s forward, despite an initial lack of physical evidence or photographs, dozens of expeditions have been mounted to the Congo to search for what has been repeatedly described by searchers as "a living dinosaur" or "the last living dinosaur."[5] These have included professional expeditions staffed by respected biologists (such as Dr. Roy Mackal from the University of Chicago), religiously inspired groups (more on them in a bit), documentary TV crews (from the British Broadcasting Corporation [BBC], National Geographic, the History Channel, etc.), and even a Kickstarter-funded expedition in 2012 (to the tune of almost $29,000!). The story of the Mokele-Mbembe continues to be regarded as true by some and has even inspired the plots of two movies, most recently 2012's *The Dinosaur Project*. Much like Nessie up in Loch Ness, the existence of the Mokele-Mbembe seems fairly implausible, yet numerous individuals have not only believed in it but spent large amounts of money and time trying to find it.

Critical Thinking About Nessie and Mokele-Mbembe

A brief review of the often-presented evidence in support of Nessie and Mokele-Mbembe shows similarities with that found for the existence of Bigfoot. For both the dinos and the man-ape, the primary support comes from eyewitness accounts and less than amazing video or photographic evidence. Physical evidence is essentially nonexistent for both Nessie up in Scotland and Mokele down in Africa, unlike that presented by Bigfoot enthusiasts. A major commonality among them, though, is the acknowledged presence of hoaxing that played a major role in getting the stories started. Remember the "Surgeon's photograph" mentioned earlier? Taken in 1934 shortly after all the excitement and publicity surrounding the monster of Loch Ness began, it became an iconic symbol and image for Nessie supporters. Combined with the early eyewitness stories, it also greatly shaped what people thought the monster looked like.

In 1975, a newspaper article first exposed that not only was the photograph faked, but explained exactly how the fakery had been accomplished (Martin & Boyd, 1999). It was a deliberate hoax perpetrated by a movie maker and big-game hunter named Marmaduke Wetherell. Wetherell had himself fallen prey to a hoax, believing he had discovered tracks made by Nessie during a "hunt" funded by the newspaper the *Daily Mail*. Rather than from a plesiosaur or monster, they turned out to be hippopotamus tracks made using a dried foot.[6] The paper then ridiculed Wetherell publicly

[5] These explorers have obviously never heard of *Denver, the Last Dinosaur*. It turns out that he's your friend, and a whole lot more.

[6] At the time, it was apparently popular to use such things to make umbrella stands. The past was pretty weird, huh?

for falling for such a stunt. Carving a fake body and head based on the description of the monster floating about in the public mind, Wetherell and his sons attached this to a toy submarine, took it to the lake and shot several photographs. Passing the film onto a friend, who then gave it to the "surgeon" it became named after, the iconic photo was subsequently sold to the *Daily Mail*, which published it and trumpeted that the Loch Ness monster had been photographed. This all came to light when his family came forward after Wetherell's death (much like the Ray Wallace faked footprints).

Other photos and videos have also been revealed as hoaxes or are too blurry to make out good details (although the "Blob Ness" monster shows up frequently in them). The original eyewitness stories aren't particularly reliable, but they did capitalize on something very prominent in the public's imagination at that time, thanks to Hollywood and a movie about a giant ape. One of the first major monster movies, *King Kong*, had opened in Scotland in the spring of 1933, just a couple of months prior to the sighting by George Spicer that would set the template for how the monster looked. In *King Kong*, there is a scene in which a creature attacks a group of humans trying to cross a lake. The creature has a long, flexible neck, a small head, and a hunched or rounded back, virtually identical to the description Spicer included in the letter that was published in the *Inverness Courier*:

> I saw the nearest approach to a dragon or prehistoric animal that I have ever seen in my life. It crossed the road about fifty hard ahead and appeared to be carrying a small lamb or animal of some kind. It seemed to have a long neck which moved up and down . . . and the body was fairly big, with a high back.

In *King Kong*, the creature then moves from the water onto the land to pursue the humans as they try to escape. In neither the film nor in Spicer's account are feet seen (in the film they are hidden by water or bushes), but in both a curving tail is seen. Finally, in the last part of the movie scene, the creature grabs a human in its mouth and moves about, foreshadowing the "animal" that Spicer reportedly saw in the mouth of his monster.

Similar to how the Patterson–Gimlin film's Bigfoot mirrored that of earlier widely recounted yet unconfirmed tales, this seems to be a case of life imitating art. Spicer undoubtedly saw something, but the similarities between his description and the scene from *King Kong* seem too close for comfort. In fact, it was only after the release of *King Kong* and Spicer's statement that anyone else saw the creature as a long-necked beast. Prior to that, *no one* described any type of long or snake-like neck on the (very rare and little reported) sightings of some "monster." As will be shown in the next section, this is not the only cryptid whose appearance appears to be taken directly from a movie, rather than real life. Still, we can't fall prey to the correlation versus causation fallacy and assume that just because two things are related that one caused the other. So we need to look for other converging lines of evidence to show that Nessie is unlikely to be a living plesiosaur . . . and there are plenty.

In addition to the problems with the origins of the story, the geological and biological obstacles to a large marine reptile living in a Scottish lake are basically insurmountable. If, in fact, there is an extant population of plesiosaurs or similar creatures living in Loch Ness, then it only holds that there must be a breeding population that has sustained them for the past 65 million years. The number needed would no doubt leave behind physical evidence of their existence, especially in a relatively closed environment such as Loch Ness. However, in the past 80 years, not a single bone or body has been found, despite all the searching. Even the fossil record shows no plesiosaur remains more recent than 65 million years old. Moreover, although huge (22 miles long, 3–4 miles wide, and as deep as 755 feet in places), Loch Ness is not anywhere big enough or rich enough in prey to support a single apex predator as large as a plesiosaur, let alone a breeding population. It's also much too cold to support any type of large, cold-blooded reptile (which plesiosaurs most likely were). In addition, the creature would have to surface to breathe air, which would result in numerous sightings and photographs of it *per day*.

Another major problem is that an actual plesiosaur could not behave in the way that Nessie is often "seen" or thought to. Perhaps the major problem is the image of a long, sinuous neck rising out of the water, which is impossible. Based on everything we can tell via their fossils, plesiosaurs would *not* have been able to move their necks in such a fashion. Instead, their necks would have gone straight out from their bodies and could have moved side to side and down, but not up in the way Nessie is often pictured (see Figure 9.3).

FIGURE 9.3 Pleisosaur.
Illustration by Ryan Long.

Where would the creatures have come from, even if they had survived the mass extinction of 65 million years ago? This is a very reasonable question, given that Loch Ness is above sea level and was landlocked until the canals built to it in the 1820s. A further problem is that prior to 10,000 years ago *Loch Ness did not exist.* It is a glacial valley and was completely covered in ice for around 2.5 million years. So a large group of the creatures would have had to have survived in some other location for 65 million years, then migrated over land into Loch Ness sometime fairly recently. Given that the folklore of the region is rich with descriptions of water-based creatures, one would think a group of huge, long-necked marine monsters would be among them if Nessie were based on an actual animal. Instead, we find many tales of water monsters, but none that resemble the Nessie of modern lore.

The story of the Mokele-Mbembe rests on even shakier evidence than that of Bigfoot or Nessie. Sauropods like the apatosaurus did not even live in swamps and rivers, although, at the turn of the 20th century when the tales of the Mokele-Mbembe began, it was thought they did. Just as with Nessie, there would need to be a large enough population of the creatures present to perpetuate the species, and such a number would leave behind the occasional carcass or bone. And yet there is no physical evidence, no decent photographs, no quality videos, and very few eyewitness accounts. The story seems to have actually spun from a couple of sentences in a book published in 1909 by Carl Hagenbeck, at the time a very well-known writer and expert on exotic animals, called *Beasts and Men*. In it, he speculates that there may be living dinosaurs in Africa, because there were native stories of huge fierce creatures that were "half elephant, half dragon." That's it. No eyewitness account or physical evidence started this story. Instead, just a few sentences in a popular book was all it took to spur the world's imagination, thanks in no small part to its fascination with the newly discovered and exhibited fossils of the mighty dinosaurs.

Perhaps the most damning piece of information in this creature's tale is that numerous times, well documented by those who believe in the creature's existence, the native tribes and peoples of the Congo region have flatly denied that it is real and insisted that it is only a story about an imaginary animal. This includes those tribe members who lived closest to where one was reportedly killed and eaten in the 1950s! In fact, the Europeans living in the area when Hagenbeck's book came out ridiculed the statement and claims, but to no avail. The story was up and running, fueled by wild speculation and, as it turns out, a somewhat hidden agenda.

Earlier it was mentioned that a number of the expeditions to find Mokele-Mbembe were religiously inspired. The most well publicized, largest expeditions to find this living dinosaur were undertaken to help find support for the beliefs of a group of evangelical Christians called young earth creationists

(YEC). These are people who believe that the world and everything in it were created in the recent past (6,000–10,000 years or so) because of their literal interpretation of the Book of Genesis. In fact, pastor Eugene Thomas and William Gibbons (who converted to Christianity during an expedition led by Thomas) are two of the most prominent Mokele-Mbembe hunters and are quite open about doing so in order to "prove" evolution wrong.[7] Milt Marcy, another YEC who funded numerous Mokele-Mbembe hunts undertaken by Gibbons, even self-published a book about the "evolutionary agenda" to remove religion from science and public life. Although now defunct, Gibbons and Marcy had even founded "Creation Generation, a creation science and dinosaur ministry" (Divine Intervention, n.d.). Not unlike Melba Ketchum, whose belief in Bigfoot led her to interpret her data questionably, the YEC Mokele-Mbembe believers use their religious belief to justify their continued search for an impossible creature and their discounting of things that go against their beliefs. Their certainty and faith in something that has not been backed by reputable evidence is constantly reinforced by confirmation biases and by ignoring both how science tests hypotheses and the methods of critically examining information.

It is important to wrap this section up by noting that there are in fact dinosaurs currently alive on Earth. Turns out they are surprisingly common, seen by billions of people worldwide, and are called birds. Yes, those birds, the same ones that defecate on your car. Even though most people think that all dinosaurs died out in the worldwide extinction event of 65 million years ago, modern birds are in fact all descended from the group of dinosaurs called theropods. So the next time you bite into that turkey sandwich, revel in the power as you tear a dinosaur apart with your bare teeth and in the knowledge that your critical-thinking skills can tear apart poor arguments for their larger cousins still being alive.

ALIEN GOAT KILLERS IN LATIN AMERICAN AND BEYOND

Unlike the other cryptids described previously, whose stories date back a century or so, the chupacabra ("goat sucker" in Spanish) is a relatively recent entry into cryptozoological lore. In the mid-1990s, stories began appearing in Puerto Rico of farm animals being killed and drained of their blood via small holes in the neck (Radford, 2011). Soon, hundreds of animal deaths (including lots of goats, hence the name) across Central and South America were being blamed on this mysterious creature. Much like with Bigfoot and Nessie, the initial reported sighting became hugely influential. A Puerto Rican woman named Madelyne Tolentino gave the first detailed report of the creature, describing it as a two-legged, long-limbed creature with dark eyes and spines along the back

[7] Their writings make it clear that this is their intent. Also made clear is that they do not actually know anything about evolutionary biology and the evidence for how natural selection over time occurs.

FIGURE 9.4 Chupacabra, as originally described in 1995.
Illustration by Ryan Long.

(Figure 9.4). For several years other eyewitnesses also described creatures that seemed to come straight out of a science fiction or horror movie. The beast was also frequently described as being around 1 meter (3–4 feet) tall and hopping like a kangaroo, but stories also described it as flying away or swooping down on prey from the sky. Sightings of creatures as far away as Russia have been described in a similar way and labeled as a "chupacabra."

One skeptical investigator (Radford, 2011) traces the entire original description of the chupacabra back to a science fiction movie that was seen by the original eyewitness not long before giving her description of the creature. Keep in mind that Tolentino's description was repeated countless times in various media forms across the years, especially early on. In the 1995 film *Species*, the main character, Sil, is a human–alien hybrid whose look was designed by Oscar-winning artist H. R. Geiger.[8] Sil walked upright but was slightly hunched over, had spikes going down her spine, large dark eyes, long thin fingers and claws, and amazing agility and jumping ability. In fact, you could basically

[8] Geiger was also responsible for the creatures from the *Alien* film series, as well as other nightmarish images in paintings and sculpture.

interchange "Sil" and "chupacabra" in the respective stories and descriptions and few people would really notice. Using our critical-thinking skills, *ruling out rival hypotheses* seems very necessary in this case. Certainly, Toletino could have seen an actual creature that coincidentally looked almost exactly like the monster from *Species* . . . but a rival hypothesis is that she just projected her memories of Sil onto the shadowy figure she reportedly saw. Indeed, when questioned about this by Radford (2011), she was recorded as saying:

> It would be a very good idea if you saw it [the movie *Species*]. The movie begins here in Puerto Rico, at the Arecibo observatory. [The monster] made my hair stand on end. It was a creature that looked like the chupacabra, with spines on its back and all . . . . The resemblance to the chupacabra was really impressive.

But remember, she saw this film *before* she "saw" the chupacabra, not after. Her description of the chupacabra looked like Sil, not the other way around.As is the case of most eyewitness accounts of Bigfoot, Nessie, and other cryptids, she was not seeing and describing what she actually saw. Instead, she described what she saw filtered through the lens of what she expected to see. Her "sighting" and description then proceeded to set the stage for what the chupacabra looks like . . . leading more people to report seeing it. This is really the same process that happens with all cryptids—people "see" what they expect to see, twisting the reality of the situation to fit their preconceived notions and beliefs (think back to Chapter 5 and that pesky confirmation bias once again).

Interestingly, the descriptions of the chupacabra began to change drastically once it "migrated" into the United States, particularly Texas and the southwestern states, in the early 2000s (Figure 9.5). Suddenly, vicious, hairless, dog-like creatures were being spotted in areas where livestock was

FIGURE 9.5 Chupacabra, as reported in the early 2000s in the United States.
Illustration by Ryan Long.

being killed and drained of blood. In what could be viewed as a watershed moment for cryptozoology, a number of these mystery creatures were even shot, killed, or trapped. At last, the crème de la crème of cryptid evidence—a physical specimen—had been obtained! Subsequent examination, both physical and genetic, confirmed that these creatures were nothing really new. All of the "chupacabras" examined have been found to be known creatures, mostly coyotes or dogs, with severe cases of mange (a disease initiated by skin mites that causes animals to lose their fur, develop odd scabs and lesions, and have a strange, rank smell).

What caused a vampiric, two-legged alien creature to turn into a mangy mutt? It seems that chupacabra became almost a catchall expression for "creature that I can't identify." Given the predatory behavior of coyotes and wild dogs, it's not surprising that the scene would be set to make someone believe in the chupacabra. Canines will often attack and kill large numbers of animals in a single herd, especially if those animals cannot escape, and they do not always eat their prey. In fact, they most often kill by biting the neck, which can cause puncture marks and bleeding out. Sound familiar from the description of the chupacabra's behavior? If this happened to you or your neighbor, and then you saw an unrecognizable animal wandering about (because, honestly, a canine with no hair and mange is a pretty odd looking critter) after hearing stories of the goat-sucker for 5 years, most people's first thought is going to go in that direction.

The media, as it has in all of the other cryptid stories in this chapter, also shares part of the blame for spreading such stories. In this case, newspapers not only uncritically reported and oversensationalized Toletino's description of the creature, but also failed to report on more likely explanations (as first described in Chapter 6). Television shows, including fiction like *The X-Files* and supposed nonfiction like *MonsterQuest*, had programs focused on the cryptid. Even today, there are monthly sightings and news stories of the chupacabra in the United States and across Latin America, and each time more people "learn" about it. The stories spread, people become primed to see something that is not actually there, and the cycle repeats itself once again.

CRYPTOZOOLOGY—SCIENCE OR PSEUDOSCIENCE?

As briefly shown here, the possibility that the cryptids reviewed in this chapter actually exist is small to nonexistent when carefully considered and examined. But despite that, many cryptozoologists will insist that they are being scientific and using science to examine the "life" of these creatures. As such, it seems a good point in this book to remind our readers how we can distinguish between scientific and pseudoscientific fields of study. As detailed extensively in Chapters 2 and 3, both sciences and pseudosciences make claims about how the world works and try to find out whether those claims are accurate. The difference is in *how* they set about doing so.

Scientists are constantly trying to disprove their own hypotheses, not confirm them, and they subject themselves to open criticism via the peer-review process. Scientists rely on replication and verification to confirm new information. This is all in direct contrast to pseudoscientists, who tend not to find a strong need of evidence before accepting a hypothesis and routinely make untestable claims. A balance of closeness and openness is crucial to being a good scientist and critical thinker (remember, have an open mind, but not so open that your brain falls out). However, it swings too far toward openness in the pseudoscientist, who often accepts any claim as long as it fits within what he wants to believe.

Supporters of cryptozoology often argue that it is a scientific and legitimate subfield of biology (e.g., Arment, 2004). However, its constant reliance on anecdotes as evidence (all of the "eyewitness testimony" supporting the existence of Bigfoot, as one example) and lack of repeatable, verifiable evidence severely undercuts such arguments. The harsh rejection of criticism is another hallmark of pseudoscience. The history of cryptid hunting shows a consistent pattern of scientists willing to consider the claims of believers and help them investigate. Unfortunately, the opinions of professional scientists (which almost inevitably go against what the cryptozoologists want to believe) are often ignored. Rather than address the skeptics' criticisms of their methodology or conclusions, most cryptozoologists will instead defend their ideas not with evidence but personal attacks on the critics. For instance, Ketchum's response to the negative reviews of her Bigfoot DNA was not to take a careful look to see whether she could make it better, but instead to start her own journal to publish it. Having been through the peer-review process for scientific publications dozens and dozens of times, your first author (CWL) can say that it can be a very long, trying process. But I will also say that each of those review processes resulted in a vastly improved manuscript, helped inform future research for myself and my colleagues, and helped the science improve. Never once did I get a negative review and then start my own journal as a result (even if I was a bit tempted to do so!).

Another major blow against cryptozoology as a science is that the discovery and cataloging of new animals already exists and is doing quite well for itself in the field of professional biology. It was mentioned at the start of this chapter how cryptozoologists do not seek out new creatures such as beetles, preferring instead the sensationalistic, legendary creatures like Bigfoot or Nessie. That is a shame, as beetles total some 30% of all living animals, with over 350,000 species known and more discovered yearly. For example, during a single expedition to Indonesia in 2013, 98 new beetle species were identified! Even larger creatures are still being discovered regularly, with over 300 new mammals being identified in the first decade of the 21st century (Conniff, 2010). However, the scientists who actually discover new creatures undergo rigorous training, often spending years studying biology before engaging in carefully planned field expeditions guided by prior knowledge and reasonable hypotheses about where the new animals might be found.

This is all in sharp contrast to the amateurs that populate cryptozoology, who are often ideologically driven and know little about fieldwork in zoology, ecology, evolution, and other critical aspects of biology. They also seem to have negative opinions of "mainstream" science in general, and often view it as somehow conspiring to make them unable to find these legendary creatures. In reality, though, finding a living dinosaur or large hominid would be the discovery of the century, earning major scientific acclaim! Scientists would be flocking to research such creatures . . . if they existed.

CONCLUSIONS

So what conclusions can the critical thinker draw from a review of the stories about and evidence presented for the existence of the cryptids discussed in this chapter? With the history of changing descriptions, exposed hoaxes, poor quality or inconclusive photos and videos, and lack of direct physical evidence, there is really only one conclusion—Bigfoot, Nessie, and the like are a product of human wishful thinking, an over reliance on our fallible memory and perceptual skills, and a media that is always looking for a new shocking story to drive readership.

This sort of careful and scientifically minded analysis, though, is extremely unlikely to sway the true believers. In an interaction that drove home this point for me (CWL), I was the moderator of a panel discussion in the fall of 2014 at a conference hosted by the campus group I advise. The panel consisted of three scientific skeptics and three believers in various paranormal phenomena. Representing the cryptozoology side of things was Carl Hartline, at the time the Associate Executive Director of the Mid-America Bigfoot Research Center. During the question and answer part of the panel, an audience member asked Hartline what type of evidence it would take for him to accept that Bigfoot was not real. After thinking for a minute, he said, "I guess that if you gave me the ability to see all the animals that are alive across North America at the same time, and I couldn't see a Bigfoot anywhere, then I would stop believing." In other words, if he were granted omniscience, then he would change his mind. Given the lack of god-like powers we are currently able to imbue humans with (smartphones notwithstanding, of course), it seems all too likely that he and most cryptid believers will keep following the wise words of the rock band Journey, namely, "Don't Stop Believing."

QUESTIONS FOR REFLECTION

1. *In 2015, a new bird species (the Sichuan bush warbler) was discovered in China. Does the discovery of a new bird species, at this stage of scientific discovery, mean we should be more open-minded regarding the existence of the chupacabra?*
2. *The Internet can play both a dampening role on cryptozoology (in giving people easy access to information that debunks Bigfoot) but can also help create a filter bubble that legitimizes the beliefs of people inside a community. How might we re-engineer the Internet to serve skeptical purposes? In other words, what would need to occur to make accurate scientific information more readily accessible than pseudoscientific information?*
3. *Cryptozoologists do sometimes rely on what seems to be a "scientific method." But they apply it inconsistently. In the case of Melba Ketchum's DNA evidence for Bigfoot, what do you consider her greatest violation of good scientific methodology?*
4. *If cryptozoological creatures were plausible, what do you think would be different about who does the research, and where it was published?*

REFERENCES

Adamnan, Ninth Abbot of the Hy Monastery. (1874). *Life of Saint Columba, Founder of Hy.* William Reeves (Ed.). Edinburgh, Scotland: Edmonston and Douglas. Retrieved from http://legacy.fordham.edu/halsall/basis/columba-e.asp

Arment, C. (2004). *Cryptozoology: Science & speculation* (1st ed.). Landsiville, PA: Coachwhip Publications.

Bigfoot Field Research Organization. (2015). *Geographic database of Bigfoot/Sasquatch sightings and reports.* Retrieved from http://www.bfro.net/gdb

Buhs, J. B. (2009). *Bigfoot: The life and times of a legend.* Chicago, IL: University of Chicago Press.

Coleman, L., & Clark, J. (1999). *Cryptozoology A to Z: The encyclopedia of loch monsters, Sasquatch, chupacabras, and other authentic mysteries of nature.* New York, NY: Simon and Schuster.

Conniff, R. (2010). *Meet the new species.* Retrieved from http://www.smithsonianmag.com/40th-anniversary/meet-the-new-species-748819/?all&no-ist

Dendle, P. (2006). Cryptozoology in the medieval and modern worlds. *Folklore, 117,* 190–206.

Divine Intervention. (n.d.). *Epidose #6: Living dinosaurs? Cryptozoologists and dinosaur hunters, Dr. William Gibbons and Milt Marcy.* Retrieved from http://divineintervention.typepad.com/divine_intervention/2008/03/episode-6-livin.html

Hill, S. (2013). *The Ketchum project: What to believe about Bigfoot DNA "science."* Retrieved from http://www.csicop.org/sb/show/the_ketchum_project_what_to_believe_about_bigfoot_dna_science

Loxton, D. , & Prothero, D. R. (2013). *Abominable science! Origins of the Yeti, Nessie, and other famous cryptids.* New York, NY: Columbia University Press.

Lozier, J. D., Aniello, P., & Hickerson, M. J. (2009). Predicting the distribution of Sasquatch in western North America: Anything goes with ecological niche modeling. *Journal of Biogeography, 36*(9), 1623–1627.

Martin, D. S., & Boyd, A. (1999). *Nessie: The surgeon's photograph exposed*. East Barnet, UK: Martin and Boyd.

Radford, B. (2011). *Tracking the chupacabra: The vampire beast in fact, fiction and folklore*. Albuquerque, NM: University of New Mexico Press.

Spicer, G. (1933, August 4). Is this the Loch Ness monster? Letter to the Editor of the *Inverness Courier.*

Sykes, B. C., Mullis, R. A., Hagenmuller, C., Melton, T. W., & Sartori, M. (2014). Genetic analysis of hair samples attributed to yet, bigfoot, and other anomalous primates. *Proceedings of the Royal Society B, 281*(1789).

Timmer, J. (2013). *Bigfoot genome paper "conclusively proves" that Sasquatch is real.* Retrieved from http://arstechnica.com/science/2013/02/bigfoot-genome-paper-conclusively-proves-that-sasquatch-is-real

CHAPTER 10

EVALUATING HEALTH CLAIMS IN ALTERNATIVE MEDICINE

No one likes to be sick, feel unwell, catch an illness, or have a disease. Most of us would (and do) go to great lengths, spending large amounts of time and huge sums of money, to avoid illness. In the United States alone, best estimates are that we spend close to $3 *trillion* per year on health care, or about $10,000 per person (World Health Organization, n.d.). Although this number will obviously vary among countries, especially those with socialized medicine, health care costs nonetheless are a significant portion of most developed countries' total spending, around 8% to 10% of gross domestic product (GDP) in western Europe and South Africa, for example.

Given the trillions and trillions of dollars spent on health care every year, it seems imperative to make sure that this money is used wisely, both for the benefit of the individual consumer and the overall health of the world's economy. An especially pertinent question from a scientifically skeptical viewpoint is this: Does this treatment work for the problem? The answer to this question underlies this chapter and the next two, where we will carefully examine the claims and evidence for a variety of treatments that fall under the umbrella term of "complementary and alternative medicine (CAM)"[1] used to treat physical and mental health problems. Before doing that, though, we need to have some solid operational definitions in place for the often-confusing litany of terms that are thrown about when discussing this topic.

DEFINING TERMS AND LEVELS OF EVIDENCE

The National Center for Complementary and Integrative Health (NCCIH) offers guidance in helping people sort out the various terms and meanings. Broadly speaking, alternative medicine and complementary medicine are those "health care approaches developed outside of mainstream Western, or conventional, medicine" (NCCIH, n.d.) Alternative practices are used in place of conventional methods, whereas complementary practices are used together

[1] Others who are a bit less generous than we are use the term "so-called alternative medicine" to refer to these practices. We will let our readers work out that acronym for themselves.

with conventional methods. More recently, many CAM proponents have begun using the term "integrative" health care to describe "conventional and complementary approaches together in a coordinated way" (NCCIH, n.d.).

Evidence-Based Practice

Contrast these definitions with that of evidence-based practice (EBP). EBP has been defined as "the conscientious, explicit, and judicious use of current best evidence in making decisions about the care of individual patients" (Sackett, Rosenberg, Gray, Haynes, & Richardson, 1996, p. 71). Another definition describes it as "healthcare practice that is based on integrating knowledge gained from the best available research evidence, clinical expertise, and patients' values and circumstances" (Dickersin, Straus, & Bero, 2007). In the real world, this often translates into using medicines, therapies, and diagnostic assessment methods that have been demonstrated to be effective via well-controlled clinical trials. Although a relatively new term, the concept of EBP dates back over 2,000 years to Hippocrates: "There are in fact two things, science and opinion; the former begets knowledge, the latter ignorance" (Hippocrates, *The Law*, Book IV, Section I, Treatise II, translation by author CL). As the father of Western medicine, Hippocrates realized that good, quality evidence was needed to be able to declare treatments effective (Singh & Ernst, 2008).

Unfortunately, most practitioners throughout history have not practiced what Hippocrates preached, instead relying on personal beliefs and anecdotes to guide their use of various medicines and techniques. But with the arrival of scientifically informed and tested medicine, EBP has "revolutionized medical practice, transforming it from an industry of charlatans and incompetents into a system of healthcare that can deliver such miracles as transplanting kidneys, removing cataracts, combating childhood diseases, eradicating smallpox and saving literally millions of lives each year" (Singh & Ernst, 2008, p. 7). Indeed, the earliest practitioners of EBP were responsible for challenging some very well-established and, in hindsight, horrendous practices, including bloodletting and ingesting mercury (a toxic heavy metal) as a cure for any illness. Physicians and nurses, such as Ignaz Semmelweis and Florence Nightingale, were key in the fight for improved sanitary conditions in hospitals by letting the data and evidence guide their practices, rather than tradition and authority.[2]

Levels of Evidence

One thing you will notice in the preceding operational definitions is that EBP refers to a method of making decisions, whereas CAM refers to a type of treatment. In other words, EBP starts with the patient and asks what is

---

[2] Sadly, despite being guided by evidence and having large amounts of data to support their ideas both were ridiculed by physicians of the day, who took great offense to their suggestions and attacked them in print and verbally. Those cognitive biases we covered in Chapter 5 have always been with us.

the best evidence for what will help to achieve a particular outcome (e.g., symptom relief or disease avoidance). As such, from the standpoint of a practitioner using EBP, there are not "alternative" or "conventional" treatments. Instead, there are three types of treatments, each with varying levels of evidence for its use or nonuse:

- Evidence-based treatments (EBTs)—those procedures, medications, and the like that have been reliably shown to cause improvement in various symptoms
- Non-EBTs—those procedures, medications, and the like that have been shown *not* to cause improvement in various symptoms
- Poorly studied treatments (PSTs)—those procedures, medications, and the like that have not been studied well enough to determine their impact on various symptoms, or for which there is conflicting evidence regarding their effectiveness

It is crucial to be aware of which treatments improve what symptoms, as not all treatments are equally effective for treating everything. Contrary to the claims of snake-oil salesmen of the 1800s and 1900s, who offered various tinctures and concoctions as a cure-all for anything that ails you, there are no treatments that will act as a panacea and fix all diseases or illnesses. Likewise, a medication or procedure that is effective at treating one problem is not guaranteed to work for others, so claims for treatment must be evaluated individually. To illustrate this point, we can take a look at antibiotics. Undoubtedly one of the triumphs of modern medicine, antibiotics of various kinds are effective at treating a wide range of problems caused by bacteria. So although they are EBTs for *Streptococcus* or *Staphylococcus* infections, antibiotics are simultaneously non-EBTs for viral infections, such as influenza or the common cold. For yet another problem, antibiotics also could be a PST. As we see in the next chapters, this is an important point to keep in mind when examining CAM.

A final aspect to consider in the discussion between EBT, non-EBT, and PST is that the categories are not static. In other words, treatments can move from "mainstream" to "alternative" and vice versa. This can lead to significant confusion, such as when a practice formerly labeled as CAM because of a lack of research on its effects is soundly examined in controlled trials and found to actually be effective. For example, the NCCIH website provides a list of "10 most common complementary health approaches among adults" for the year 2012. Topping the list is "natural products" (vitamins and supplements), followed by "deep breathing." But as we will see in the next chapter, there are a number of herbal supplements that have been repeatedly found to be useful for the treatment of specific problems. Likewise, the impact of "deep breathing" (often called diaphragmatic breathing) has been well studied and found to help improve pulmonary functioning in asthmatics and to be a good stress reducer. So deep breathing is an EBT for several problems, and as such could be considered in treatment plans developed by a practitioner using EBP for those specific problems.

Changing Names and Kinds

This confusion among EBT, non-EBT, and PST has led to significant shifts in how certain governmental agencies in the United States discuss CAM. For instance, the National Institutes of Health (NIH)-sponsored NCCIH was formed in 1991 as the "Office of Alternative Medicine (OAM)," changed its name to the "National Center for Complementary and Alternative Medicine (NCCAM)" in 1998, and then changed it again to the current "National Center for Complementary and Integrative Health" in 2014. This appears to be, at least in part, due to the negative associations people may have with the term "alternative medicine." Indeed, on the current version of their website it is actually difficult to find the word "alternative" used by the organization to discuss any of the practices described on their site, which is a major shift from the past. In fact, they say that "True alternative medicine is uncommon. Most people who use nonmainstream approaches use them along with conventional treatments" (NCCIH, n.d.).

What is not detailed, though, is why NCCIH continues to describe treatments that should fall into the non-EBT category (for example, acupuncture or homeopathy) as being in the PST or even EBT categories, while describing EBT procedures or methods as being "complementary." As Australian performer Tim Minchin says in his brilliant beat poem *Storm*, "By definition (I begin), alternative medicine (I continue) has either not been proved to work or been proved not to work. Do you know what they call alternative medicine that's been proved to work? Medicine" (https://youtu.be/HhGuXCuDb1U). This conflation of various treatments with varying levels of evidence, lumping them together as "CAM" does no one any good, particularly those seeking effective health care options.

This naming shift and confusion isn't the only change that NCCIH has made, though. During it's time as OAM and NCCAM, the organization listed five main types of CAM that it was studying or supporting research for, which were:

- Whole medical systems (e.g., homeopathy, naturopathy, Ayurvedic medicine, and traditional Chinese medicine, Bach flower remedies)
- Mind–body medicine (e.g., meditation, prayer, art therapy, music therapy)
- Biologically based practices (e.g., herbal and dietary supplements)
- Manipulative and body-based practices (e.g., chiropractic, massage, craniosacral therapy)
- Energy therapies (e.g., Reiki, acupuncture, therapeutic touch, electromagnetic therapy)

With the change in name, though, came a change in grouping, moving from five to two categories and a catch-all:

- Natural products (e.g., herbs, vitamins, minerals, probiotics, and other dietary supplements)
- Mind and body practices (e.g., yoga, chiropractic manipulation, massage, meditation, acupuncture, relaxation, hypnotherapy, movement therapies)
- Other complementary health approaches (includes naturopathy, homeopathy, traditional Chinese medicine, Ayurvedic medicine, and anything else that doesn't fit in the preceding two categories)

However they are grouped, the discerning scientifically minded physician, clinician, or other health care provider (and you, dear reader, by the end of this book) will see that there is a mix of EBT, non-EBT, and PST in the NCCIH's groupings and discussion of CAM. Treatments that have been well studied and supported (such as relaxation techniques) are placed side by side with both highly pseudoscientific non-EBTs (such as acupuncture, which we will discuss) as well as PSTs (such as probiotics[3]). Given this conflation, we find it more effective to talk about the level of evidence available to support a given treatment for a given condition, rather than just lump medicine and therapy into CAM or conventional categories.

GLOBAL USE OF CAM

Over the last 70 or so years, medical science has made enormous strides in improving the health of the planet's population. From the worldwide eradication of smallpox, which has saved an estimated 5 million lives annually, to the almost total erasure of deaths in developed countries from diseases like diphtheria, measles, mumps, rubella, and polio (United Nation's Children's Fund [UNICEF], n.d.), the success of scientifically guided medicine is undeniable. And that is just from the successful implementation of vaccines and routine immunizations! The major causes of death in developed countries have shifted massively over the past 100 years, moving from infectious diseases to "lifestyle" diseases such as heart disease or cancer. People have a longer life span than at any point in history, thanks in no small part to medical breakthroughs such as antibiotics, innovative surgeries, organ transplants, and more. It would seem that trust in "conventional" medicine should be exceedingly high, that the public would be clamoring for increasing amounts of it compared to any other form of treatment.

[3] Although there is an increasingly large amount of both basic and clinical research examining the impact of the microorganisms living in our gut with our physical and mental health, many of the health claims for ingestion of probiotic supplements are at this point not well supported.

Interestingly, though, during this same time period, there has been a steady rise in the use of CAM across the globe. Public acceptance and use of various non-EBTs or PSTs that are grouped under the CAM umbrella appear to have steadily increased since the 1970s, from an average of 14% of people in 1970 to over 32% by the 2010s (Frass et al., 2012). The most commonly used specific treatments worldwide are chiropractic manipulation, homeopathy, herbal medicines, acupuncture, and massage (most of the surveys did not include questions on "energy medicine," so the rates of their usage are not well known).

Such high use rates aren't cheap, either. Most recent findings put out-of-pocket costs for CAM in the United States at around $34 billion per year (NIH, 2009). Add that to the estimates of $7 billion spent in the United Kingdom each year, the $3.1 billion in Australia, the $320 million per year in South Africa, or the billions upon billions spent in other countries on treatments that have little to no evidence supporting their use (Mpinga et al., 2013) and it begins to look like a huge amount of money could be put to much better use. So (one might ask), why would people spend so much money on things that don't work? Wouldn't they realize these treatments were ineffectual and turn to something else for help? The answer to those questions is a bit complicated, but as we see in the next section, it boils down to "because they do work, just not for the reasons people think they do."

PLACEBOS, REGRESSIONS, AND THEIR EFFECTS

Physicians and other health care providers who use EBP rely on the findings from valid and reliable research studies (think back to Chapter 2) to decide what is most likely to work in the treatment of a patient's problems. Two of the main reasons that good research is so critical to the development of EBTs is (a) how easily bias can creep into our everyday decision making (as discussed in Chapter 5) and (b) how influenced we are by powerful social forces, such as advertising (as discussed in Chapter 6). But a particularly salient third reason why strongly controlled research is needed in health care is because of something called the placebo effect.

The Placebo Effect

As we have repeatedly seen throughout this book, people's beliefs can have a powerful impact on how they process the information that they are exposed to. But belief, it turns out, can have not just a major effect on one's mind, but also on one's body. A placebo can refer to any type of sham or inactive medical treatment or procedure. The most commonly used placebos are "sugar pills" (pills with no actual medication in them, just fillers) or fake infusions (an injection that contains only sterile water), although there have been placebo surgeries and other procedures used in various research. The placebo effect is defined as "the measurable, observable, or felt improvement in health or behavior not attributable to a medication or invasive treatment that has been administered" (Carroll, 2015). In the real world, this would mean that someone would be unknowingly given a false treatment (thinking it is the real thing) and then show improvement *even though he or she has not gotten any active treatment*. It's the grown-up equivalent of a parent kissing your scraped knee and telling you that makes it feel better, so you stop crying because the pain has gone down.

Admittedly, this sounds ridiculous at first, but decades of research have shown that placebos (Latin for "I will please") can in fact have numerous powerful effects (Goldacre, 2010). Although these effects seem to be more pronounced in subjective (the patient reporting on if they feel better) rather than objective tests (blood tests), they have been demonstrated across a wide range of conditions. Most susceptible to the placebo effect appear to be pain (acute and chronic), depression, asthma, sleep problems, and irritable bowel syndrome. To take asthma as one example, people using placebo inhalers did not have an objectively measured increase in lung functioning, but nonetheless reported (subjectively) that they could breathe more easily after using the sham inhaler.

There are a number of reasons, both biological and psychological, why people would respond to a placebo despite the lack of active ingredients. On the biological side of things, there have been a number of studies that have shown that taking placebos can change what is happening in the brain and body (Benedetti, 2014). This includes causing the production of endogenous (or naturally occurring) cannabinoids and opioids[4] as well as numerous other brain changes, from increased activation in the prefrontal cortex to the active release of dopamine. In other words, there are measurable physiological changes that, in many cases, match those changes that occur when given the real, active treatment.

Psychologically, the expectations and perceptions of the person taking the placebo become very important. When you take a drug or have a procedure done, you have expectations about how this will impact your functioning. These expectations then color your perception of how you feel and impact your behavior. Take, for example, alcohol. Even if you have never actually imbibed yourself, you likely have expectations for how people behave when they are intoxicated. Researchers over the past 40 years have found that drinking what you are told is alcohol, even when it is actually nonalcoholic, causes you to "act" drunk. For example, people act more aggressively, show more interest in erotic or violent books and videos,

[4] That's right, your brain naturally produces chemicals similar to what you ingest from marijuana and morphine. That's the entire reason ingesting those substances causes the effects they do—because we already have receptors for chemically similar things in our brain. Your brain is like your own personal drug dealer!

become more sexually aroused, and even have memory problems (Assefi & Garry, 2002). One's expectations about the effects of drinking alcohol cause large changes in one's perceptions and behavior.

This is further illustrated by contrasting the placebo effect with what's called the *nocebo effect*. You can take an inert pill and, if told that it reduces pain, you will feel less pain when administered a small electric shock. However, you could be given the exact same inert pill and be told that it increases pain, and you will then report feeling more pain when administered a small electric shock. This is the nocebo effect—when something negative happens or you feel worse after receiving a sham treatment because you expect to feel worse.

There is also increasing work showing that not just the treatment, but how you deliver the treatment can have a major impact on the strength of the placebo effect. For instance, Ted Kaptchuck, a professor at Harvard Medical School, has shown that placebos given to patients by researchers who are friendly, comforting, and show interest in the patient's personal life work much better than the same placebo given in an abrupt way with little interaction. This "care effect" appears to greatly enhance the strength of a placebo. You can further enhance it just by changing physical properties of the placebo. Giving a placebo in a capsule rather than as a pill appears to have a stronger effect. Giving the same placebo as an injection produces a still stronger effect. Other work has shown that the more expensive a placebo is, the better it appears to work; plainly packaged placebos work less well than ones in fancy boxes. Color also has an impact—red pills work better as stimulants and blue pills work better as depressants (as long as your culture associates red with activity and blue with relaxation). Presentation matters in most areas of life, and it appears to matter an enormous amount when it comes to placebos (Goldacre, 2008).

Regression to the Mean in Health

In addition to the placebo effect, a particularly frequent way that bias can creep into our decision making about health care is something we discussed in Chapter 5, called regression to the mean (RTM). In technical terms, "RTM is a statistical phenomenon that occurs when repeated measurements are made on the same subject or unit of observation" (Barnett, van der Pols, & Dobson, 2004, p. 215). Plainly stated, RTM refers to the idea that when a measurement shows something having moved to an extreme degree in one direction, it will most likely move back (regress) toward "normal" (the mean) across repeated measurements. To use a health care example, say that you have a headache: It starts small (not far from the mean of "no headache"), but then builds over time until it becomes unbearable (an extreme value from the mean). At that point, you take some type of treatment (aspirin, let's say).

Soon, you feel better and have less of a headache (a regression toward the mean).

In this example, there are three potential explanations for why your headache went away. It could be that taking the aspirin truly had the effect of decreasing your pain levels. Or it could be that the aspirin had a placebo effect because you have been conditioned to feel less pain after taking medicine and expect that to be the case. Or it could be that the pain level would have decreased regardless of what you did, and that you are confusing correlation with causation—that's RTM at work in a health situation. Without properly controlled study any one of these explanations is plausible.

So research has shown that placebos can change your physiology, your behavior, and your cognitions. It's no wonder, then, that the placebo effect needs to be very carefully controlled for when conducting studies to examine whether or not a particular treatment works. We also need to control for the tendency of problematic symptoms to regress across time and become better with no intervention before we can say that a particular treatment works. In fact, almost all treatments can "work," by which we mean "cause an effect." What we have to ask ourselves, though, is not "does this treatment work?" but instead "does this treatment work better than a placebo?" and "would this condition naturally improve over time, even with no intervention?"

Those questions are the keys, it turns out, to examining so much in the realm of alternative and complementary medicine. When no good research exists, or the research of equal quality is contradictory, we have a PST. If something reliably works better than a placebo and/or the healing effects of time, then it can truly be said to have an active effect and thus falls into the realm of EBT. When, instead, well-controlled studies find that a treatment is no better than a placebo control, it moves into the non-EBT category. This type of viewpoint will let us examine the benefits or failings of a particular type of CAM, rather than just dismiss all of the treatments that sometimes fall under that term.

THE BLIND RESEARCHING THE BLIND

Because of the placebo effect and the phenomenon of RTM, studies that examine treatment outcomes need to be very carefully designed so that we don't see a relationship when one is not actually there. This typically means that we need randomized, placebo-controlled, *double-blind* procedures in place to control for these effects. It is only by relying on high-quality clinical trials of these kinds that we begin to truly understand which treatments do and do not have an evidence base. But what does this look like in the real world?

In randomized trials, you divide the entire group of people participating in the study randomly into two groups—the treatment and control groups. This is done to even out potential differences (gender, education level, symptom severity, comorbid problems, etc.) between the groups, making them more homogenous, or similar. The placebo-controlled aspect is fairly self-explanatory: you compare the new treatment (or the old treatment for a new problem) to a placebo, rather than nothing. Placebos should generally be matched in type to the active treatments for the best control. If your treatment is in pill form, the placebo needs to be a pill as well; if the treatment is a surgical procedure, then the placebo control needs to be surgical (in terms of making incisions on the body and stitching or closing them up, even if that is all that is done). To help control for regression effects, the placebos should last as long as the active treatment being studied.

In blinded studies, the trial participants (the people who have the condition being treated) are divided into two groups—active treatment and placebo control. They are "blinded" because they are not aware of which group they are in. In this way, you can control for the expectancy aspects of the placebo effect. However, in order to control for bias on the part of the treatment provider, truly excellent studies are double-blinded: neither the patient nor the researcher knows who is in the treatment rather than placebo group. In this way, the researchers will not be able to subtly bias the results by acting differently toward one group or the other.

When evaluating the evidence about a treatment's effectiveness, whether CAM or conventional, this kind of randomized, placebo-controlled, double-blinded study is the gold standard. Studies that fail to meet these criteria are highly susceptible to bias and placebo effects, resulting in findings that show treatments to be effective when they actually are not. As we see in the next two chapters, these biases are critical when determining whether a CAM treatment is EBT, non-EBT, or PST for a particular condition.

WHY IS CAM SO POPULAR?

As shown previously, complementary, alternative, or integrative health treatments are utilized by a very large number of people globally, even when they have ready access to more conventional approaches that are likely to be more effective for most problems. So why would you go with CAM, and perhaps choose something that's known to be a non-EBT? Although the answer is multifaceted and differs among individuals, philosopher and scientific skeptic Robert Todd Carroll (2003) has collated a number of the most commonly seen reasons.

Top of the list is that there are no drugs or surgery in CAM. Many people have heard horror stories of surgeries gone wrong, or run from modern medication when they hear the litany of side effects that it may cause.

This potential for harm can make many people seek a "natural" alternative, which they think seems safer. For example, rather than take conventional medications, it might be seen as safer to take "vegetable pills," which may in turn have reportedly miraculous effects, such as limb regrowth (Figure 10.1)! Of course, you'd also have to make sure you weren't taking too much of a good thing, or otherwise terrible side effects might occur (Figure 10.2).

Along with that, many types of CAM (such as homeopathy, supplements, or essential oils) may be cheaper and easier to access than traditional

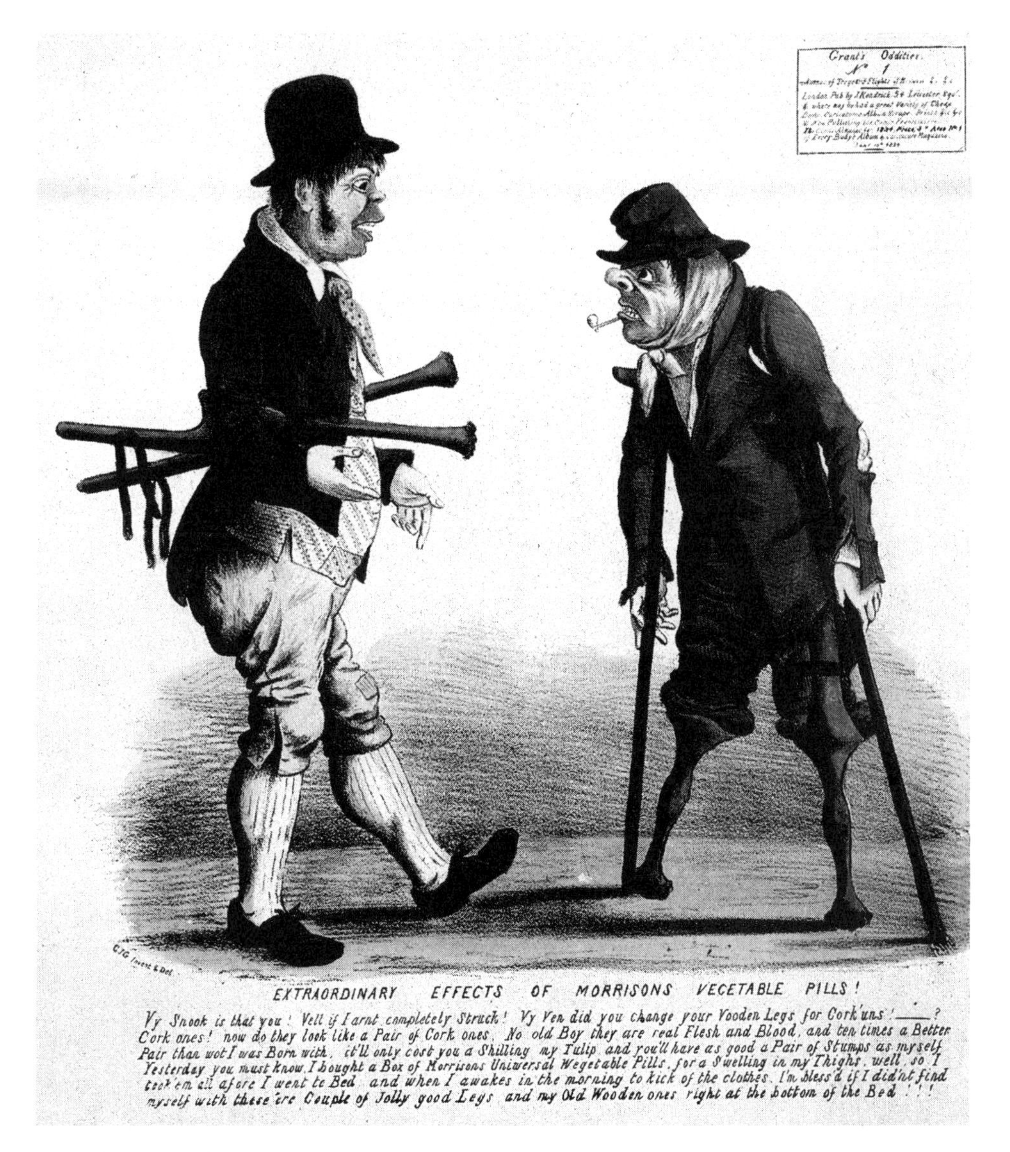

FIGURE 10.1 Extraordinary effects of Morrison's vegetable pills; severed legs made whole again.

Source: Courtesy of the Wellcome Library. Wellcome image V0011126. Color lithograph, 1834, by C. J. Grant, published by J. Kendrick, London. Reproduced under Creative Commons Attributions Only license (http://creativecommons.org/licenses/by/4.0).

FIGURE 10.2 "Singular effects of the universal vegetable pills on a greengrocer! A fact!".

Source: Courtesy of the Wellcome Library. Wellcome image V0011125. Color lithograph, 1831, by C. J. Grant, published by J. Kendrick, London. Reproduced under Creative Commons Attributions Only license (http://creativecommons.org/licenses/by/4.0).

medicine, which involves visiting a physician, paying your co-pay, getting a prescription, going to the pharmacy, paying another co-pay, and then finally going home to take your medication. This is especially salient in those countries (like the United States) without guaranteed access to medical care via socialized medicine, where one of the major causes of personal bankruptcy is excessive medical bills.

Turning to CAM can also be seen as a direct reaction to some of the failings of conventional medicine. For example, think about the last physician visit you had. For many of us, visiting a physician is not a pleasant experience. We make an appointment weeks in advance, sit in the waiting room long past the time when we are supposed to be seen, have a rushed whirlwind of an encounter with the physician, and get sent home with a treatment that is likely to work but is not guaranteed to be effective. This is especially true with chronic or poorly understood conditions, when conventional physicians may not be able to uncover the cause (or causes) of your pain or discomfort. Many CAM practitioners, on the other hand, will spend extended periods of time with you as a patient, listening intently, providing reassurance and simple answers about what you need to do to get better.

Politics also comes into play for why people accept CAM as legitimate (even if research on a particular type shows otherwise). Many CAM practitioner groups have successfully lobbied to obtain governmental licensure and regulation. For instance, at the time of this writing all 50 U.S. states have licensure and regulation boards for chiropractors, 44 license practitioners to perform acupuncture, and 17 have license naturopaths. This provides a patina of approval for those practices, as states also license physicians, psychologists, dentists, and other conventional treatment providers. Combine this with a well-funded NCCIH under the umbrella of the U.S. federal government's NIH and the implicit authentication and seal of approval become more convincing.

Another common reason is a misunderstanding of how science works. Over the past 100 years, we have seen enormous changes in how physicians and other health care providers deliver treatments, and what kind of treatments they deliver. Certainly, the history of scientific medical and mental health treatment is filled with things that are now known to be ineffective (prefrontal lobotomies, thalidomide, insulin shock therapy, stenting for stable coronary disease). Conventional health care is fallible, yes, but it is (like all of science) self-correcting. The reason we stopped using certain drugs or treatments is because we tested them and found them to not work! This stands in sharp contrast to most CAM, which offers simple answers to complex problems and does not change over time. For many people, the certainty offered by CAM practitioners is highly appealing.

Finally, many people continue to utilize CAM because, for them, it works! At least, it appears to work, resulting in pain or other symptom relief after treatment. What most people don't realize is how powerful the placebo effect can be, and that what they are paying for and why they are experiencing relief is not a result of their treatment, but a result of their belief in the treatment. Even in cases in which people are utilizing both conventional and CAM treatments (in the case of cancer, for example), they often attribute any success to CAM because of their faith in it working. As we discussed in detail in Chapter 6, you really can't trust your brain sometimes, and people seeking out CAM and extolling its benefits, even when the data show it not to be useful, is an excellent example.

TIPS FOR AVOIDING NON-EBP IN HEALTH CARE

Evidence-based medications and treatments stand on their own scientific merit, whereas non-EBT often has to deceive people into purchasing and using it. Along with employing all your critical-thinking skills, here is a handy guide to avoiding being scammed by people peddling health cures that are less than evidence based. The Federal Trade Commission's Bureau of Consumer Protection (1999) provided a list of six typical phrases and techniques used to draw people into believing false claims about treatments:

- The product is advertised as a quick and effective cure—all for a wide range of ailments.
- The promoters use words like *scientific breakthrough, miraculous cure, exclusive product, secret ingredient,* or *ancient remedy.*
- The text is written in "medicalese"—impressive-sounding terminology to disguise a lack of good science.
- The promoter claims the government, the medical profession, or research scientists have conspired to suppress the product.
- The advertisement includes undocumented case histories claiming amazing results.
- The product is advertised as available from only one source.

It is especially good to be on your guard against non-EBT if you fit into a particular demographic. Research shows that the "typical" CAM user tends to be middle-aged, female, and of higher than average education and income. The CAM user is also likely to have multiple medical conditions, although these may or may not be serious in nature (Bishop & Lewith, 2010). Many less than ethical or non–evidence-based practitioners target those who have conditions that conventional medicine often fails at treating, such as multiple sclerosis, diabetes, obesity, dementia, chronic pain, depression, and various types of cancer. These are people who are desperate to find relief and often willing to try anything. Although we all want simple solutions and definitive answers to our health problems, the reality is that physical and mental health are complex issues that often have complex, complicated diagnoses and treatments. If you find yourself fitting into these categories, keep your critical-thinking toolbox close at hand.

CONCLUSIONS

As Thomas Paine, U.S. constitutional author and founding father, once wrote: "To argue with a person who has renounced the use of reason is like administering medicine to the dead" (Paine, 1778). Between this chapter and the next two, we hope that any treatment you administer or take in the future will be evidence based rather than the alternative. Because of space constraints, we cannot cover every single type of CAM that a person might encounter throughout a lifetime. Instead, in the next two chapters we take a look at some of the most popular examples of CAM that claim to help with physical or mental health problems. These will be used to help illustrate some of the points discussed in this chapter and in the first half of the book, showing how to think critically about any CAM practices you might encounter.

QUESTIONS FOR REFLECTION

1. *Given that producers of pharmaceuticals can (and do) make plenty of money from patented medicines, would you think alternative medicine more or less plausible, in general?*
2. *What do you think is the most likely explanation for the popularity of alternative medicine? Have you ever used CAM? What was your reason?*
3. *What does the nocebo effect tell us about how humans respond to medical interventions?*
4. *What is our best scientific methodology for demonstrating the efficacy of alternative medicine?*

REFERENCES

Assefi, S. L., & Garry, M. (2002). Absolut memory distortions: Alcohol placebos influence the misinformation effect. *Psychological Science, 14*(1), 77–80.

Barnett, A. G., van der Pols, J. C., & Dobson, A. J. (2004). Regression to the mean: What it is and how to deal with it. *International Journal of Epidemiology, 34*(1), 215–220.

Benedetti, F. (2014). *Placebo effects: Understanding the mechanisms in health and disease* (2nd ed.). Oxford, UK: Oxford University Press.

Bishop, F. L., & Lewith, G. T. (2010). Who uses CAM? A narrative review of demographic characteristics and health factors associated with CAM use. *Evidence-Based Complementary and Alternative Medicine: eCAM, 7*(1), 11–28.

Carroll, R. M. (2003). *The skeptic's dictionary: A collection of strange beliefs, amusing deceptions, and dangerous delusions.* Hoboken, NJ: John Wiley & Sons.

Carroll, R. M. (2015). *Placebo effect.* Retrieved from http://skepdic.com/placebo.html

Dickersin, K., Straus, S. E., & Bero, L. A. (2007). Evidence based medicine: Increasing, not dictating, choice. *British Medical Journal, 334*(Suppl.), s10–s11.

Federal Trade Commission. (1999). *"Operation cure all" targets internet health fraud.* Retrieved from https://www.ftc.gov/news-events/press-releases/1999/06/operation-cureall-targets-internet-health-fraud

Frass, M., Strassl, R. P., Friehs, H., Mullner, J., Kundi, M., & Kaye, A. D. (2012). Use and acceptance of complementary and alternative medicine among the general population and medical personnel: A systematic review. *Ochsner Journal, 12,* 45–56.

Goldacre, B. (2008). *Bad science: Quacks, hacks, and big pharma flacks*. London, UK: Faber & Faber.

Goldacre, B. (2010). *Bad science: Quacks, hacks, and big pharma flacks*. London, UK: Faber & Faber.

Mpinga, E. K., Kandolo, T., Verloo, H., Zacharie Bukonda, N. K., Kandala, N-B., & Chastonay, P. (2013). Traditional/alternative medicines and the right to health: Key elements for a convention on global health. *Health and Human Rights Journal, 15*(1), 44–57.

National Center for Complementary and Integrative Health. (n.d.). *Complementary, alternative, or integrative health: What's in a name?* Retrieved from https://nccih.nih.gov/health/integrative-health

National Institutes of Health. (2009). *Americans spent $33.9 billion out-of-pocket on complementary and alternative medicine*. Retrieved from https://nccih.nih.gov/news/2009/073009.htm

Paine, T. (1778). *The American crisis*. Retrieved from http://constitution.org/tp/amercrisis06.htm

Sackett, D. L., Rosenberg, W. M., Gray, J. A. M., Haynes, R. B., & Richardson, W. S. (1996). Evidence based medicine: What it is and what it isn't. *British Medical Journal, 312*, 71–72.

Singh, S., & Ernst, E. (2008). *Trick or treatment? Alternative medicine on trial*. London, UK: Bantam Press.

United Nation's Children's Fund (UNICEF). (n.d.). *Vaccines bring 7 diseases under control*. Retrieved from http://www.unicef.org/pon96/hevaccin.htm

World Health Organization. (n.d.). *Global health observatory data repository*. Retrieved from http://apps.who.int/gho/data/?theme=home

CHAPTER 11

ALTERNATIVE MEDICINE FOR PHYSICAL HEALTH

In this book, you've already learned how to differentiate science from pseudoscience, how to think critically in broad strokes and about several specific subjects, and how your own brain and the world at large sometimes conspire to make you misinterpret information. In Chapter 10, we specifically presented you with reasons why it is crucial to be able to critically examine health claims to know whether a particular treatment is evidence based, non–evidence based, or poorly studied for a particular problem. This chapter places a skeptical eye on some of the more commonly utilized complementary and alternative medicines (CAMs) that are purported to cure various physical health problems. We cover some of the history and background of the treatments then delve into what the research says about their effectiveness for various problems. In this chapter we use the same three levels of evidence introduced in Chapter 10:

- Evidence-based treatments (EBTs)—those procedures, medications, and the like that have been reliably shown to cause improvement in various symptoms
- Non–evidence-based treatments (non-EBTs)—those procedures, medications, and the like that have been reliably shown *not* to cause improvement in various symptoms
- Poorly studied treatments (PSTs)—those procedures, medications, and the like that have not been well-studied enough to determine their impact on various symptoms, or for which there is conflicting evidence regarding their effectiveness

First, we cover chiropractic manipulations of the spine, which are often used not only for neck and back problems, but a myriad of physical ailments. Then, we shift to the ancient art of acupuncture, in which needles stuck into your skin in specific locations reportedly help to increase health. Next, we move to the pills and remedies of homeopathy, followed by vitamins and herbal supplements. We then end the chapter by taking a look at so-called "energy therapies" and the biological therapies of vitamins and supplements.

CHIROPRACTIC MANIPULATIONS

One of the most widely utilized treatments falling under the CAM label is chiropractic manipulation. Interestingly, many people do not know that it is a nonconventional type of treatment and assume that chiropractors are medical physicians who just specialize in the spine and back. This is probably in part the result of the ubiquity of chiropractors in many developed countries. In the United States, for instance, there are currently over 77,000 chiropractors and 40,000 chiropractic assistants (American Chiropractic Association [ACA], 2015). Although the United States has by far the largest number, Canada can count over 8,000 chiropractors in its boundaries, with another 3,000 or so in Australia, 2,000 in the United Kingdom, and smaller numbers across the rest of the world.

Another likely reason for some confusion about the status of chiropractic is that, in the United States, its practitioners go by the title "doctor," after graduating with a "doctor of chiropractic" degree. All of these degrees are granted from one of the 18 privately owned and run programs or colleges that are accredited by the aptly named Council on Chiropractic Education. Chiropractors tend to assist with this confusion by referring to those who went to medical school, graduated with an MD, and are able to prescribe medications and perform surgeries as "allopathic physicians" and themselves as "chiropractic physicians," subtly implying that the two types of training are equivalent. One website took this comparison further by stating that a "medical doctor, doctor of chiropractic, or doctor of osteopathic medicine is a practitioner of the healing arts who examines patients, analyzes the results of laboratory tests, diagnoses and treats the patient's medical condition, and advises the patient about methods of preventive health care," adding that all should be considered primary care providers (Spine Health, 2015).

History and Background

The history of chiropractic is quite interesting, but little known to the general public. Although the ACA attempts to trace the roots of chiropractic care back to ancient China and Greece, the reality is that chiropractic is relatively young, having started in 1895. In what appears to be some deliberate ignoring of history and facts, here is how the ACA's (n.d.) website describes the birth of chiropractic:

> In the United States, the practice of spinal manipulation began gaining momentum in the late nineteenth century. In 1895, Daniel David Palmer founded the Chiropractic profession in Davenport, Iowa. Palmer was well read in medical journals of his time and had great knowledge of the developments that were occurring

> throughout the world regarding anatomy and physiology. In 1897, Daniel David Palmer went on to begin the Palmer School of Chiropractic, which has continued to be one of the most prominent chiropractic colleges in the nation.

The previous paragraph leaves out a few key pieces of information. For instance, Palmer was a teacher and grocer with no medical or scientific training who, in 1886, opened up an office of "magnetic healing" in Davenport.[1] In 1895, he performed the first chiropractic adjustment, although Palmer's recounting of the event differs significantly from that of the family of the man it allegedly happened to.

In his account (D. D. Palmer, 1910), Palmer had carefully listened to the medical history of Harvey Lillard, who had partial hearing loss. Lillard had felt a popping in his back at the time he began having hearing problems some years before. Palmer then examined the man and found a lump on his back that seemed to indicate that Lillard's spine was out of alignment. Palmer then adjusted his back, causing remarkable improvement in his hearing. And thus began chiropractic treatments.

Or maybe not, if you believe the account of Lillard's daughter. In her version of the story, her father was talking to some friends outside of Palmer's magnetic therapy office. Lillard was telling a joke when Palmer came out and joined them. At the joke's punch line, Palmer slapped Lillard heartily on the back. Lillard then reported to Palmer after a few days that his hearing seemed to be a bit better.[2] According to Lillard, it was at this point that Palmer began to think about spinal manipulation as part of his magnetic healing practice and the two discussed going into business together to promote it. Although that didn't happen, 2 years later Palmer had coined the term "chiropractic" and opened the Palmer School of Chiropractic to teach others his techniques.

Also left out of the ACA's short history are the ideas about the causes of illness that Palmer promulgated and that continue to be believed and adopted by the vast majority of chiropractors. He thought that health problems occurred (any health problems, mind you) as a result of small misalignments of the spine that were later called subluxations.[3] According

---

[1] A local paper published an article at the time, saying "A crank on magnetism has a crazy notion that he can cure the sick and crippled with his magnetic hands. His victims are the weak-minded, ignorant and superstitious, those foolish people who have been sick for years and have become tired of the regular physician and want health by the short-cut method . . . he has certainly profited by the ignorance of his victims . . . . His increase in business shows what can be done in Davenport, even by a quack" (as quoted in Colquhoun, 2008, p. 6).

[2] It's important to note that there is no possible way that a neck or spinal adjustment could impact one's hearing, based on everything that we currently understand about neurology. None of the spinal nerves, for example, extend into the brain, let alone the areas of the brain involved in hearing.

[3] This can lead to confusion, as the term "subluxation" in modern medical terms means only a partial dislocation of an organ or joint in the body. These are visible via medical imaging, such as x-rays or MRIs. This is very different from the chiropractic meaning of the term.

to Palmer, these subluxations inhibited the flow of one's life energy, which he called "innate intelligence," and caused 95% of health problems. Figures 11.1 and 11.2 come from an early manual of spinal manipulation and show some supposed subluxations and that all problems can come from a subluxation in a specific spot (Forster, 1915). Spinal manipulation as used by chiropractors would then unblock this flow of energy, restoring health. No need to take our word for it, though. Here is B. J. Palmer, son and heir to the founder, who said:

> Chiropractors have found in every disease that is supposed to be contagious, a cause in the spine. In the spinal column we will find a subluxation that corresponds to every type of disease. . . . There is no contagious disease. . . . There is no infection. (B. J. Palmer, 1909, as quoted in Campbell, Busse, & Injeyan, 2000)

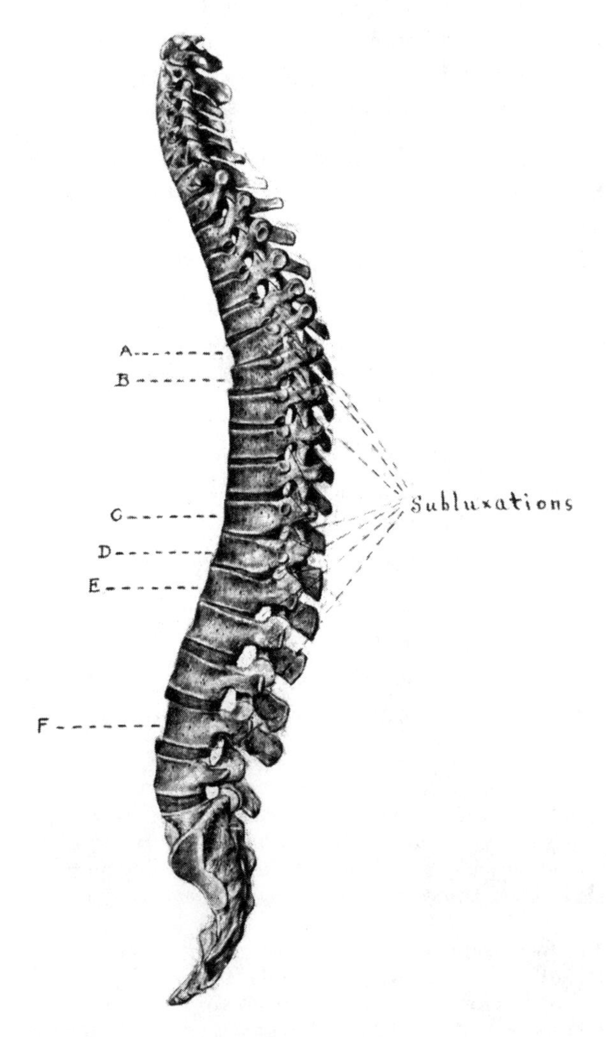

FIGURE 11.1 Subluxation
Source: Forster, A. L. (1915).

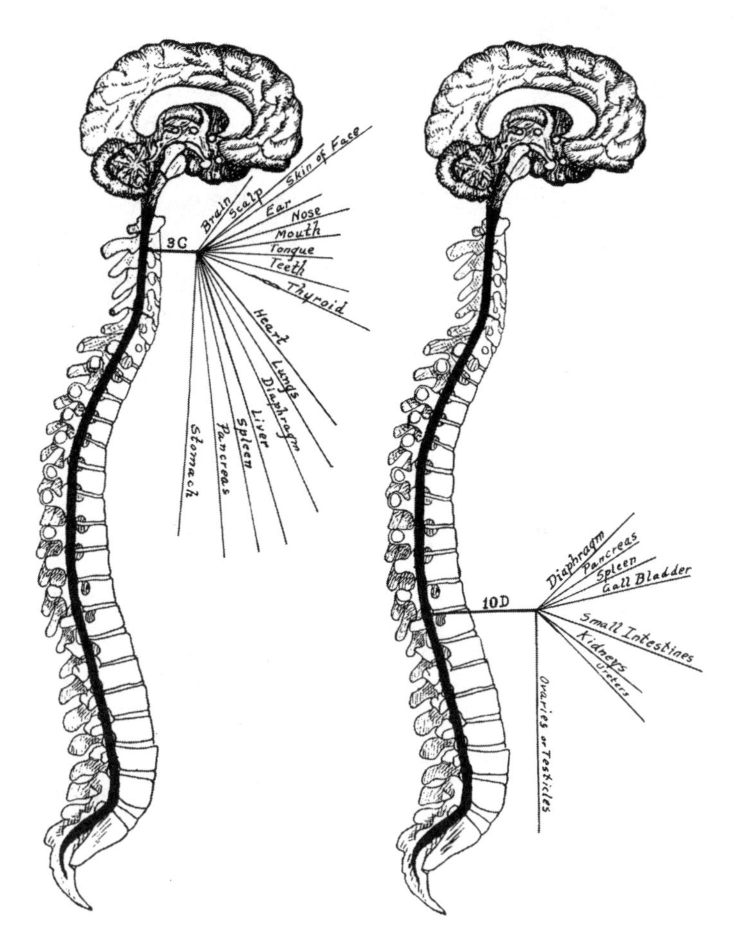

FIGURE 11.2 Subluxation problems
Source: Forster, A. L. (1915).

Remember, this was written at a time when the germ theory of disease (in which bacteria, viruses, and other microorganisms cause illness) was already very well supported through the scientific and medical discoveries of people like Louis Pasteur and Robert Koch. We already knew that certain diseases were contagious, we knew why they were contagious, and intervention efforts involving sanitation had already shown how they could be at least partially controlled. Chiropractors, however, thought it was all a load of bunk and instead followed the word of their founder.[4]

The history of chiropractic, then, is filled with dubious claims and pseudoscientific ideas about disease. But what of modern chiropractic? Just as medical science is not the same as it was in the 1890s, surely chiropractic has evolved and changed, right? Although the language has been updated

[4] In an interesting aside, Palmer's own writings refer to chiropractic as a science, art, philosophy, and religion. He saw himself as being similar to the founders of other religions in that way, comparing himself to Jesus Christ, Mohammed, Joseph Smith, and Martin Luther. These ideas would later lead his son, B. J., to consider incorporating chiropractic as a religion when it came under fire from medical groups for allegedly practicing medicine without proper training or licensure.

quite a bit since the early 1900s, the concept of subluxations is still at the core of almost all chiropractic practice (National Board of Chiropractic Examiners, 2014). There is a bit of division, though between "straights" and "mixers." The straights (a minority of chiropractors) still strictly follow Palmer's ideas on subluxation and see it as the cause of most physical (and mental) health problems. They tend to do only spinal adjustments, generally by hand alone. Mixers, on the other hand, tend to think that although disease can develop from other causes, this generally only happens when subluxations lower your body's ability to fight off invaders like bacteria or viruses. These, the majority of chiropractors, will often use a variety of approaches to healing people, including many of the CAM treatments discussed in this chapter. That both groups still rely heavily on subluxations is highly troubling from a scientific point of view, given the complete lack of evidence that such chiropractic subluxations are associated with any problem of health functioning or, honestly, that they even exist (Mirtz, Morgan, Wyatt, & Greene, 2009).

Although some modern chiropractors reject the idea of subluxations and instead focus on scientifically informed treatments, they are a tiny minority (P. H. Long, 2013). To quote Novella (2009):

> So chiropractors cannot realign the spine to fix imaginary subluxations and restore the flow of nonexistent innate intelligence. Subluxation theory is pure pseudoscience, like homeopathy or therapeutic touch, and has no place in a 21st century scientific health care system.

Research on Chiropractic

Despite this very shaky theoretical ground, there is nonetheless a fairly large amount of research into the potential benefits of chiropractic manipulations. Returning to our discussion a few sections back about CAM and levels of evidence, what can we say about the research evidence supporting (or not) the use of chiropractic manipulations for specific types of conditions? After all, an estimated 27 million people in the United States are treated by a chiropractor each year and routinely report high levels of satisfaction, so there must be something to it. Luckily, good research can tell us whether that "something" is a placebo effect or whether it is an actual, active treatment.

As mentioned previously, labeling a treatment as CAM or conventional tells us very little about how well a particular treatment works for a particular problem. Therefore, we need to examine whether chiropractic works for specific health issues rather than if it just "works." One way to determine this is to see what it is advertised to help with and look at the evidence for those problems. For example, one of your authors (CWL) performed a survey (admittedly, an unscientific one) of the websites of chiropractors operating within the city where the University of Central Oklahoma (where he teaches) is located, to see what they claim to be able to treat. Although all of them listed back pain as something they could help treat, the range of conditions listed across the websites was enormous and included carpal tunnel, fibromyalgia, osteoporosis, headaches, scoliosis, sciatica, pregnancy (not the condition of being pregnant, but instead having an "optimal" pregnancy), jaw pain, addictions, allergies, asthma, attention deficit hyperactivity disorder (ADHD), bed wetting, infertility, ulcers, thyroid problems, improving eyesight, colic, psychological trauma, shin splints, plantar fasciitis, and so on. When any one treatment is reputed to be able to cure such an enormous number of conditions, one needs to keep in mind our critical thinking (CT) maxim of "Extraordinary claims require extraordinary evidence." So what does the evidence say?

The good news for chiropractors is that large-scale research analyses (e.g., Ernst & Canter, 2006; Merepeza, 2014) are supportive of manipulative therapy being helpful in treating uncomplicated (meaning no pinched nerves or herniated discs), chronic lower back pain. It doesn't, however, appear to be more helpful than typical low-cost conventional treatments, such as nonsteroidal anti-inflammatory medications (NSAIDs, like aspirin, ibuprofen, or naproxen) or exercise. None of them are highly effective, but all offer mild relief for chronic low back pain. It should be noted, though, that manipulative therapy of the kind often included in these reviews is not only practiced by chiropractors, but also by physical therapists and physiatrists.

For everything else on the prior list, and in all of the literature using well-designed and controlled research, the results are not good news for chiropractors. Aside from a mild benefit for low back pain, there is no good evidence that chiropractic manipulations are effective at treating or curing any other problem. Literally, nothing else. Large-scale reviews find no significant effect of chiropractic on headaches (Bronfort et al., 2004), asthma (Kaminskyj, Frazier, Johnstone, & Gleberzon, 2010), ADHD (Karpouzis, Bonello, & Pollard, 2010), bedwetting (Huang, Shu, Huang, & Cheuck, 2011), or anything else outside of low back pain. In other words, using our terminology developed in the previous chapter, chiropractic would be qualified as an EBT for chronic lower back pain, but as a non-EBT for everything else.

Given the lack of logical, rational, or scientific reasons that could explain why adjusting one's spine would help with lung functioning, enuresis, or the like, this is not very surprising (Singh & Ernst, 2008). What is surprising, though, is the continued advertisement of chiropractic being useful in the treatment of these conditions, which is not confined to our little previous website survey, but is instead a widespread problem across the globe (Ernst & Gilbey, 2010). Also disturbing is that so many people are unaware of the non-EBT basis for the vast majority of what chiropractors claim to treat, resulting in huge amounts of money being spent on care that is no better than placebos. For example, in 2012 the U.S. government, via Medicare, spent $496 million for chiropractic treatment, which does not even include what was paid by people out of pocket or through private insurance.

Another infrequently discussed issue, particularly by chiropractors, is the very real risk associated with chiropractic manipulations. Particularly dangerous are neck manipulations, which have been used since the beginning of chiropractic (Figure 11.3). Several studies in the past 15 years have found individuals at a greatly increased risk of vertebral artery dissection (VAD). A VAD is a tear of the inner lining of the artery that supplies blood to the brain, which causes blood to clot on the arterial wall, obstructing blood flow. This in turn can result in head and neck pain, stroke symptoms, and myriad other problems. Case-controlled studies have found large increases in chances of a VAD occurring within 1 week (Rothwell, Bondy, & Williams, 2001; Smith et al., 2003) to 1 month (Dittrich et al., 2007) of a neck manipulation. The greatest increase in risk appears to be in persons who are younger than 45 years old. Because of these studies and others, the American Heart Association and American Stroke Association released a joint statement warning practitioners of the risk of cervical artery dissections (including VADs) from neck manipulations and urging them to inform patients of these risks (Biller et al., 2014).

In chiropractic, then, we have some very questionable beginnings, emphasizing belief over evidence, followed by a century's worth of anecdotal evidence that it works. Millions of people[5] are treated by chiropractors each

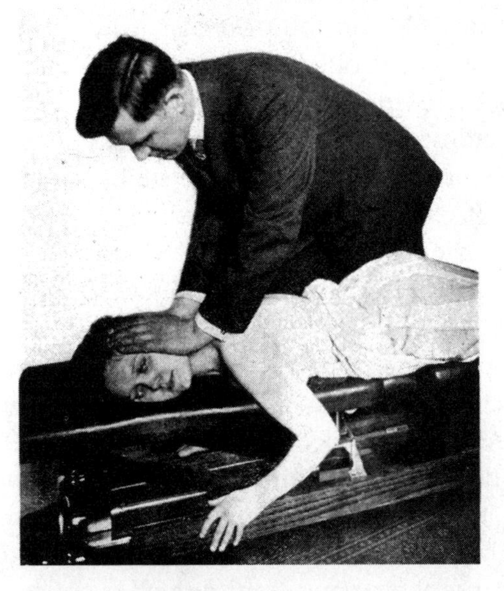

FIGURE 11.3 Neck manipulations
Source: Forster, A. L. (1915).

[5] It must be mentioned that today there are even some veterinarians who claim to practice chiropractic on nonhuman animals, particularly horses.

year, for a wide range of problems, and yet the good data we have suggest it should be considered an EBT *only* for uncomplicated, chronic lower back pain. Combined with the relative risks, high costs, and low benefits compared to other treatments, it is even questionable to use it for that. There is no support for it being useful for any other condition, despite the claims of its practitioners. As such, the vast majority of chiropractic practice would need to be considered a pseudoscience and a non-evidence-based CAM treatment.

ACUPUNCTURE

We all have a friend who cannot stand to go to the physician's office because he is afraid he will have to get a shot (or perhaps that's you!). The idea of sticking a needle into his flesh gives your friend an upset stomach, makes him sweat, or even get woozy and pass out. But what if your friend could overcome almost any illness simply by having needles stuck into the appropriate part of his body? This is the claim of those who practice acupuncture, a CAM treatment that has gone from obscure to mainstream since the 1970s.

Typically regarded as one component of traditional Chinese medicine (TCM), acupuncture use globally has exploded over the past 40 years. Today, it is estimated that well over 6% of Americans have seen or currently see an acupuncturist, and the rate is even higher among Europeans and Australians (Zhang, Lao, Chen, & Ceballos, 2012). Estimates in East Asian countries are higher still, where approximately a quarter of Japanese and Taiwanese having used it at some point in their lives, and with acupuncture being the most widely used CAM in China each year. There are currently about 30,000 licensed acupuncturists in the United States, and most countries now have regulatory boards or agencies for it. Despite its rise in popularity, most private insurance plans in the United States (and most other countries) will not cover the cost of acupuncture treatments.

History and Background

The history of acupuncture is very interesting and shows how influential stories and politics, rather than evidence, can be in the popularization of a health treatment (Singh & Ernst, 2008). Although the origins are shrouded in mystery (with some claiming it was used in Europe over 5,000 years ago and others that it was invented around that time in China), our earliest records show that it was in use by at least the 2nd century BCE in China. In fact, most of the tenets of modern acupuncture were already established then: the idea that a "life force" or energy called *Ch'i* (or *Qi* or *Chi*, pronounced "chee") flows through your body via channels called meridians, which are connected to your various internal organs; that blocks in one's *Ch'i* causes diseases or illness; and that placing needles into the meridians can unblock the flow of *Ch'i*. Figure 11.4 shows an actual acupuncture chart from the 1700s displaying the supposed meridians and where the needles should come from.

Although there was a smattering of interest in acupuncture outside China and East Asia prior to the 20th century, the use died out by the mid-1800s across Europe and the Americas. Even in China, where it had been used for thousands of years, it fell into disuse as people became more interested in modern medical knowledge and advances. The entire reason acupuncture experienced a revival in China was because of a single political force—the Chinese Communist Party. Party Chairman Mao Tse-Tung pushed for the reintroduction and widespread use of TCM, including acupuncture and herbal medicine, across the country. This push was caused not by any evidence supporting the use of TCM, but instead to help the Communist Party fulfill promises of accessible health care in a country with very few modern medical practitioners and a huge, rural population.[6] TCM was provided not by trained physicians or nurses, but by "barefoot doctors"—basically farmers with a small amount of

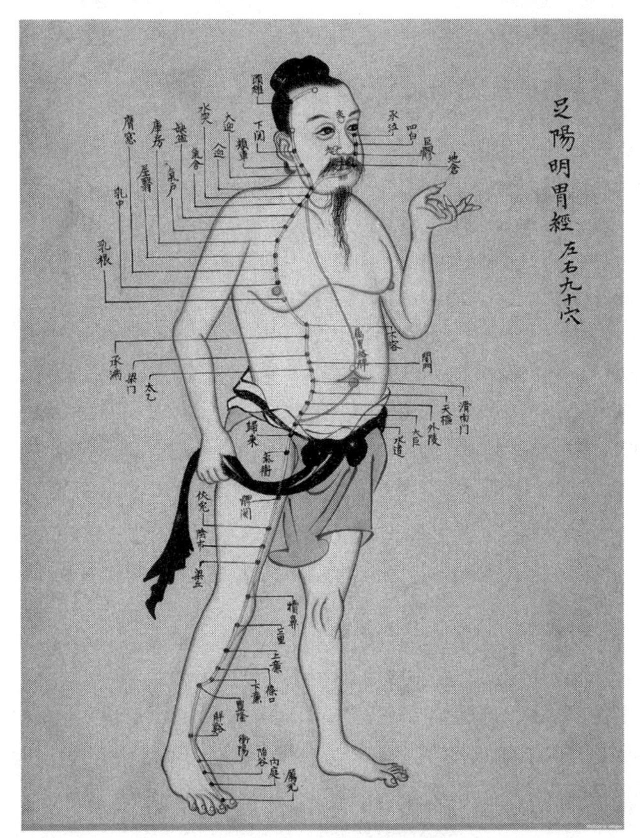

FIGURE 11.4 Acupuncture points.
Source: Courtesy of the Wellcome Library. Wellcome image V0018499. Reproduced under Creative Commons Attributions Only license (http://creativecommons.org/licenses/by/4.0).

[6] In a testament to how political a move this was, Chairman Mao was quoted as having said, "Even though I believe we should promote Chinese medicine, I personally do not believe in it. I don't take Chinese medicine" (Li, 2011, p. 84).

medical training who combined Western and traditional Chinese treatments. It wasn't for another two decades that there would be much notice of this movement outside of China, a result of its isolationist policies.

The trip by U.S. President Richard Nixon to China in 1972 was historic for a number of reasons, helping break down a 25-year wall of separation between Cold War enemies. For the purposes of this book, it also served to introduce the plethora of treatments that made up traditional Chinese health care. Nixon and his companions were shown a patient having surgery while awake, using only acupuncture rather than medical anesthesia. This, combined with a *New York Times* reporter writing on the effectiveness of acupuncture for his pain after a surgical procedure performed in Beijing in 1971, helped spark enough interest that the first legal center for administering acupuncture in the United States was established in 1972 (Colquhoun & Novella, 2013).

Unfortunately, left out of what Nixon saw and other claims circulating at that time about using acupuncture as the sole anesthesia in open-heart surgeries, was that it was all a lie. The patients in these surgeries were actually given a combination of analgesics, morphine, and other painkillers and *then* had the acupuncture needles placed in, for showmanship (Singh & Ernst, 2008). This deception, however, did not become well known until much later. By the time it was known, acupuncture had already begun its spread into Europe, the Americas, and elsewhere, and today is one of the most commonly used CAM treatments in the world. As such, it ranks with chiropractic as being one of the most widely studied CAMs, which helps us to answer the main question for this section of the chapter: Is acupuncture evidence based, non–evidence based, or PST?

Research on Acupuncture

As a result of its growing popularity, a large amount of research was conducted looking at the effect of acupuncture for a variety of conditions in the 1970s. In 1979, after a fairly good number of clinical trials, the World Health Organization (WHO) issued a report stating that acupuncture had been shown to be useful for about 20 conditions, listing problems as varied as asthma, headaches, diarrhea, and the common cold as having responded well to acupuncture. This was a huge boon for the practitioners, who were apparently vindicated by scientific evidence. Of course, the question that was still puzzling was how exactly it worked, because most Western physicians and scientists recognized that the concept of *Ch'i* was highly pseudoscientific. Researchers were eager to find out why acupuncture worked, because it apparently did. This led to the next phase of research into acupuncture, which turned out not so well for those practicing it.

As outlined previously, the main principles on which acupuncture is based are exceedingly strange, but no stranger than those found in other prescientific views on health (think about the Innate Intelligence of chiropractic described previously, or the Law of Similars in homeopathy, which will be detailed later). Researchers began putting forth hypotheses about how acupuncture could work, based on our current understanding of biology. Unfortunately, they did not get very far with this before coming upon a pair of rival hypotheses—that the early positive results could very well be the result of two factors, neither of which was taken into account in the pioneering studies of the 1970s.

In fact, much of the early, positive research into the effectiveness of acupuncture suffered from two major flaws that undermined the results. These two factors—the placebo effect and regression to the mean (RTM; both heavily discussed in our previous chapter)—were not controlled for because studies were small, nonrandomized, not blinded, and had no placebo controls. As discussed in Chapter 10, the placebo effect can be very powerful, and the more novel, dramatic, and expensive a placebo is, the better it works. When, in the 1990s, funding was made available to perform more clinical trials of acupuncture, they were designed to meet much higher standards than prior studies. This included the use of various kinds of "sham acupuncture" as the placebo controls—with needles either only barely inserted (not far enough to hit the "meridian" points) or misplaced from their traditional points. More recently, several groups developed an even better placebo—a telescopic needle that looked and felt as if it was piercing the skin, but in fact did not puncture the skin at all. This improved placebo allowed the highest quality trials to date of acupuncture's effectiveness.

In contrast to early studies, the trials of acupuncture conducted in the 1990s and 2000s were not overwhelmingly supportive. Once the placebo effect and RTM were controlled for, effects were either nonexistent or so small as to be clinically insignificant. In fact, high-quality sham and real acupuncture consistently show equal effectiveness in treating various types of chronic pain—in other words, all the impact of acupuncture seems to be the result of the placebo effect (Ernst, Lee, & Choi, 2011). Although there have been a number of positive clinical trials during this time, they tend to have relatively large problems in their study design and analysis or far overstate their conclusions. Many of these trials also compare a standard sham acupuncture to acupuncture with electrical stimulation, meaning that their placebo control is not adequate. On the whole, we have to agree with the Oxford Centre for Evidence-Based Medicine, whose summary was "What we are left with, though, is a conclusion that acupuncture is no better than a toothpick" (Cherkin et al., 2009). In short, acupuncture is a non-EBT for any and all conditions.

Another major problem, aside from the lack of evidence, is that the hypotheses underlying acupuncture are pseudoscientific. There is no evidence of any sort of measurable "life force" like the concept of *Ch'i*, that this force moves through the body via certain meridian points, or that it can become blocked and then unblocked via the insertion of needles into the skin. Not only are the central ideas of acupuncture nonsense, but the clinical trial results and expert consensus is that it just does not work. In other words: "The best controlled studies show a clear pattern, with acupuncture the outcome does not depend on needle location or even needle insertion. Since these variables are those that define acupuncture, the only sensible conclusion is that acupuncture does not work" (Colquhoun & Novella, 2013, p. 1362).

HOMEOPATHY

Homeopathy is a practice that is older than chiropractic and younger than acupuncture, but like both is a frequently used type of CAM. Homeopathy was developed in Germany and then spread globally, but how the practice is regulated and endorsed varies widely. For instance, Belgium only allows physicians, dentists, and midwives to practice homeopathy; South Africa tightly regulates the practice and requires the equivalent of a master's degree to be a "homeopathic practitioner." Meanwhile, homeopathic remedies can be bought over the counter in pharmacies in the United States and Germany. Rates of use vary widely, from around 2.3% of the U.S. population who spend over $3 billion per year (National Center for Complementary and Integrative Health [NCCIH], 2009) to about 10% in the United Kingdom, spending around $50 million. In India, where a reported 100 million people rely solely on homeopathy for their medical care (Prasad, 2007), there are almost as many homeopaths as there are medical physicians (Roy, Gupta, & Ghosh, 2015).

History and Background

The origins of homeopathy are well understood and documented, unlike those of acupuncture. In the late 1700s and early 1800s, German physician Samuel Hahnemann (Figure 11.5) was dissatisfied with the medical practice of his day (which was, to say the least, barbaric in many ways and non-evidence based). Via experimentation on himself and lots of thought, he came up with what he called the "Law of Similars." According to Hahnemann, those substances that caused a symptom in a healthy person could be used to cure the same symptom in an unhealthy person. For instance, because onions can cause your eyes to water and your nose to drip, a person with allergies can take onion to cure those same symptoms. This "like cures like" idea caused him to produce *remedies* to treat various illnesses through careful observation of what substances caused which symptoms in healthy people.

Now, on the surface, this idea of "like cures like" seems a bit similar to what modern medicine does in terms of vaccines and immunizations. After all, the measles, mumps, and rubella (MMR) vaccine is a mixture of live,

FIGURE 11.5 Portrait of Samuel Hahnemann
Source: Courtesy of the Wellcome Library. Wellcome image L0047458. Portrait by Marmaduke B. Sampson, published by S. Highley, London. Reproduced under Creative Commons Attributions Only license (http://creativecommons.org/licenses/by/4.0).

attenuated viruses that you inject into your bloodstream to confer protection against those three diseases. Isn't that a case of "like curing like?" If this were all that there was to homeopathy, the answer would be "yes, they are similar" (even though vaccines use the actual causal factors of the diseases, rather than some sort of analog). However, Hahnemann did not stop with the Law of Similars, and instead further posited that if one were to dilute a remedy (generally in ethanol or water), it would simultaneously increase the strength of the cure and decrease the side effects. Further, the potency of the remedy could then be increased more if you shake and then strike the remedies after the dilution, called succussion. By 1820, Hahnemann had coined the term *homeopathy* (from the Greek for "similar suffering") and published numerous volumes on various homeopathic remedies and what they treated. These books are still consulted by modern homeopaths and have a revered status in the profession.

The dilutions recommended by Hahnemann (and still used by modern homeopaths) were, to put it mildly, a bit astounding. He recommended diluting a substance by a factor of 100 per each stage of succussion (that is, one part of the original substance to 99 parts water, and thereafter one part of the new mixture to 99 parts water). Hahnemann stated that a 30C dilution (that is, 30 stages of dilution and succussion) was generally the optimal mixture. After this 30th stage, the actual homeopathic pills are produced by taking an inert pill (most often made of lactose) and placing a single drop of the solution on it. But at this level of dilution, the chance that even a single molecule of the original substance will be in a homeopathic pill is infinitesimal—approximately one in a billion billion billion billion (Singh & Ernst, 2008). Stated plainly, homeopathic pills are basically guaranteed to have no active ingredients at all.[7] This is why they are not similar to vaccines, all of which contain small concentrations (but still on the order of billions) of the antigen or virus that they help protect against. These antigens or viruses essentially prime the body's immune system to be able to fight off large-scale amounts of the antigen or virus if exposed to it in the future. But in homeopathy there is nothing present to cause any sort of biological response.

So if there is nothing of the original substance in the homeopathic pills, how could it cause any sort of symptom relief? Hahnemann was not particularly bothered by the idea of dilutions being able to go on forever, as he knew nothing at the time about atomic theory and molecules being the smallest portion of a chemical. Modern homeopaths make claims that water somehow retains a "memory" of the original substance, so that even though there are no molecules of the substance, its healing properties are still in place. The idea of water having a memory, or retaining the properties of things it comes into contact with, is patently absurd. Why would the memory be limited to homeopathic substances, for example, and not the raw sewage it once came into contact with, or the pharmaceuticals you took with a glass of water last night? The dilution aspects are also mind-boggling. Physicist Robert Park calculated that to produce a 30C remedy one would need to take one molecule of the active substance and dissolve it in a container of water approximately 30 billion times larger than the Earth. Stated plainly, both of homeopathy's main tenets are in direct conflict with everything we know about biology, physics, and medicine (e.g., atomic structure, the law of mass action, and dose–response relationships). There is no scientific evidence to back up the extraordinary claims of Hahnemann and his followers about why homeopathy should work.

Research on Homeopathy

But despite the complete lack of plausibility and evidence to support the main ideas behind homeopathy, there have still been a large number of clinical trials conducted to examine if homeopathy does, perhaps, work. Much like we have seen in acupuncture already, small, poorly designed studies that were susceptible to bias (either because of poor randomization or lack of blinding) tended to show larger effects for homeopathic remedies across a variety

[7] In an arresting example of the amount of dilution we are talking about in homeopathy, a commonly sold remedy for influenza is called *Oscillococcinum* and is most often sold at a 200C dilution. With approximately 10^{80} molecules in the known universe, a 40C dilution would result in one molecule of the original substance in the entire universe. The 200C dilution, therefore, would require 10^{320} *more universes* to guarantee one molecule would be present (Park, 2008).

of conditions. Because of the large number of such trials, meta-analyses that included them tended to show a small but significant effect of homeopathy compared to placebo. However, larger, placebo-controlled clinical trials repeatedly show that homeopathy does not reliably work better than placebo, especially when compared to conventional treatments for similar conditions (Linde et al., 1999; Shang et al., 2005). A large-scale review of systematic reviews (sort of a meta-meta-analysis) came to the same conclusion, that "the most reliable evidence . . . fails to demonstrate that homeopathic medicines have effects beyond placebo" (Ernst, 2010, p. 460). A more recent, well-funded review of 176 studies conducted by the Australian government came to the same conclusions, writing that:

> There was no reliable evidence from research in humans that homeopathy was effective for treating the range of health conditions considered: no good-quality, well-designed studies with enough participants for a meaningful result reported either that homeopathy caused greater health improvements than placebo, or caused health improvements equal to those of another treatment. (National Health and Medical Research Council, 2015, p. 6)

Homeopathy is full of ideas that are in direct odds with our knowledge of physics and biology but fit nicely with a prescientific view of the world and healing. Despite being used by hundreds of millions of people worldwide, reviews of several decades of high-quality evidence show that it is no more beneficial than placebos for any condition. Not only is it implausible, it just doesn't work. These conclusions also spill over into other systems of CAM that spring from the ideas of homeopathy, such as Bach's flower remedies, homeopathic "vaccines" made using isopathy, or the use of homeopathy by veterinarians. Our conclusion that homeopathy is a non-EBT for any condition is in line with the review by the UK's House of Commons Science and Technology Committee in 2010:

> To maintain patient trust, choice and safety, the Government should not endorse the use of placebo treatments, including homeopathy. Homeopathy should not be funded [by the government] and the [government] should stop licensing homeopathic products.

THERAPEUTIC TOUCH AND OTHER ENERGY THERAPIES

A favorite concept employed by many different CAM therapies is that of "energy." In chiropractic, the life force or energy is what is supposedly blocked by subluxations; in acupuncture the needles release the *Ch'i* energy so it can do it's magical healing through the body. Even in Samuel

Hahnemann's original writings on homeopathy he refers to the purpose of homeopathy as restoring a "vital force" or spirit to cure diseases. The next set of CAM therapies we look at, the energy therapies, take this idea of some type of nonmaterialistic "life energy" even further, proposing that it can be directly manipulated without having to rely on spinal adjustments or needles or remedies, as long as you are properly trained. Some examples of commonly practiced energy therapies include therapeutic touch (TT), Reiki, biofield energy healing, crystal healing, and qigong. The healing claims of these run the full gamut of conditions, from making wounds heal quicker and decreasing pain to improving carpal tunnel syndrome and causing weight gain in premature infants.

Energy therapies (sometimes called energy medicine) are highly diverse in name and practice, but all share two underlying beliefs. First is that disruptions to the human body's "energy field" can cause innumerable diseases and/or disrupt the healing process. As one example, the website of the Therapeutic Touch International Association (TTIA) (n.d.) says:

> Therapeutic Touch is based on the idea that human beings are energy in the form of a field. When you are healthy, that energy is freely flowing and balanced. In contrast, disease is a condition of energy imbalance or disorder. The human energy field extends beyond the level of the skin, and the Therapeutic Touch practitioner attunes him or herself to that energy using the hands as sensors.

In Reiki, which was developed in Japan in the 1920s, reference is made to the manipulation of one's *ki*, analogous to the *Ch'i* life force described in Chinese culture. Reiki means "universal life force" or "spiritually guided life force energy" (depending on your translator) and when this life force "is low, then we are more likely to get sick or feel stress, and if it is high, we are more capable of being happy and healthy" (International Center for Reiki Training, n.d.).

The second underlying belief common to energy therapies is that the "energy field" (whatever its name) can be manipulated in some way by trained individuals. These manipulations can help to restore its balance, thus curing disease or speeding up the healing process. Practitioners of TT (which claims to be an evidence-based therapy on their international association's website) do this by holding their hands a few inches above the body of the sick person[8] and moving their hands rhythmically to help with "facilitating the symmetrical flow of energy through the field . . . energy may be transferred where there is a deficit or energy may be mobilized or repatterned from areas of congestion" (TTIA, n.d.). Reiki practitioners may either lay hands directly on their clients

---

[8] Yes, you read that correctly. People who do TT do not actually touch you. A bit of deceptive naming, some might say.

or place their hands slightly above the body area where they are reportedly directing the Reiki energy for "rebalancing."

Given the varieties of names, ideas, and history behind each type of energy healing, space constraints prevent us from examining them as thoroughly as some of the other practices in this chapter. However, we can critically examine the two beliefs that undergird the entirety of this field of CAMs to see whether there is any plausibility or evidence to support such claims. As mentioned previously, the idea of some type of "energy" has been used for thousands of years when it comes to healing practices. But what, exactly is "energy?"

In modern physics, the word *energy* primarily refers to the ability of a system to do work. "Work" in this case is the application of a force along the distance an object travels (force multiplied by distance), meaning that work causes a change or movement in matter. Therefore, energy is a property or condition of objects, one that (consistent with the laws of thermodynamics) can be transferred or converted into other forms, rather than an object itself. The energy of an object, then, is how much work the object can do on some other object.[9] There are a number of different types of energy, including chemical energy, mechanical energy, nuclear energy, electrical energy, thermal energy (heat), and radiant energy (light).

Each form of energy can be converted into other kinds, although the process can be complex. For example, plants convert solar energy into chemical energy via the process of photosynthesis. Animals then eat the plants and transfer the chemical energy stored in them by breaking those down into various molecules for their own body to use. Photovoltaic solar cells absorb light from the sun and convert it into electrical energy, which we can then use to power our cell phones, cars, homes, and more. Energy is, quite literally, all around us, as all matter stores energy of some type. Remember Albert Einstein's famous equation $E = mc^2$? That allows us to calculate the amount of energy inherently stored in resting matter, which is its mass multiplied by the speed of light squared. In other words, there is a huge amount of energy in very small amounts of matter, if you just know how to convert and transfer it.

The practitioners of various "energy therapies" are not quite as specific (or backed up by centuries of dedicated scientific research) as physicists and chemists. Instead of observable, measurable forms of energy (such as chemical or electrical), energy in CAM typically refers to some type of immaterial, immeasurable "force." These types of "energy" are often rooted in prescientific and even religious views of the world, with the terms *energy*, *spirit*, and *soul* being interchangeable for some practitioners. For instance, the

[9] This is a very simplified explanation of energy as defined in physics and chemistry, leaving out lots of exciting details, but it will be enough for our purposes. Interested readers can check their local library for introductory physics textbooks, which go into detail about these types of processes.

Ch'i and *ki* concepts found in Chinese and Japanese cultures typically refer to some sort of vital, universal life force that flows through all living matter. One (of the many) larger definitions is:

> Qi is the basic material of all that exists. It animates life and furnishes functional power of events. Qi is the root of the human body; its quality and movement determine human health. There is a normal or healthy amount of qi in every person, and health manifests in its balance and harmony, its moderation and smoothness of flow. (Kohn, 2005, p. 3)

However, different practitioners believe different things about this life force or energy, some that it is material and part of our world, others that it comes from a spiritual plane of existence and is thus supernatural. In other words, "energy" in CAM circles in no way means the same thing that it does in the scientific world.

Another major difference between CAM and scientific definitions of "energy" is how they are measured. Physicists and other scientists typically use *joules* to measure the amount of energy in an object or situation. A joule is carefully described as the amount of energy needed to produce 1 watt of power for 1 second and forms the basis for measurement of energy (and energy potential). Anyone, anywhere, can measure how many joules are contained in something or in a situation using certain kinds of equations. Contrast this to how CAM "energy" is measured, which is purely via self-report. The practitioners claim to be able to sense or detect the life force, biofield, or other types of so-called energy, which then allow them to manipulate it. This claim, though, can be put to the scientific test, and it was, by a 9-year-old girl named Emily Rosa.

Research on Energy Therapies

In 1996, Emily saw a video of people performing TT. These people claimed to be able to physically feel what they called the "human energy field" (HEF) and then used their hands to manipulate it in order to improve health. Impressed but a bit skeptical, she devised a clever yet simple experiment for her 4th grade science fair in order to test whether or not TT practitioners could actually feel this energy coming off of the human body. What she did was conduct a single-blind experiment, in which she had the TT practitioners place both of their hands through a screen made from a typical science fair display board so that the practitioner could not see on the other side. Emily then placed one of her hands over one of the TT practitioner's hands, well within the range in which the practitioner could supposedly detect an HEF, and say whether her hand was over the practitioner's right or left hand. A simple protocol, and one that (if they could actually physically feel this hypothetical HEF) should have been very easy to pass.

Unfortunately for TT, none of the practitioners got a better than chance level of correct answers, only guessing the position of Emily's hand correctly at an average of a little over 4 out of 10 trials. With the help of her mother (a nurse), father (a statistician), and a physician, Emily conducted more trials after her science fair, submitting the resulting experiment for publication. To date, Emily is the youngest person to ever author a published, peer-reviewed journal article (Rosa, Rosa, Sarner, & Barrett, 1998), and in no less than one of the world's top medical journals (the *Journal of the American Medical Association*).

Emily applied one of the key tools of CT—falsifiability—to an extraordinary claim and showed, simply stated, that it was nonsense. Other researchers tested the replicability of her findings, confirming that Emily's initial results were accurate (R. Long & Bernhardt, 1999). This type of plausibility testing—or examining whether the underlying beliefs of a CAM therapy are true—can literally dismantle all the claims of TT (and many other energy therapy) practitioners. After all, if a person cannot reliably and accurately detect whether he or she is in contact with an HEF (or aura or whatever word they use), how can they possibly hope to manipulate it? It would be the equivalent of a physician attempting to do surgery without being able to tell when or if the scalpel was actually entering the body of the patient.

To put a further damper on the potential usefulness of TT, the authors of a large-scale review found that there was no evidence that TT helps promote the healing of wounds (O'Mathúna & Ashford, 2014), and another review found that the vast majority of published research on "nontouch biofield therapy" (energy therapies with no physical contact, like TT) were of such poor quality that it was impossible to make recommendations about their effectiveness (Hammerschlag, Marx, & Aickin, 2014). The results are similar for other energy therapies, such as Reiki: The evidence either shows no benefit or the studies are poorly controlled. All energy therapies studied to date are non-EBT for any condition.

The biggest problem we noticed across the research literature was the lack of placebo controls. Instead, most studies would compare the energy therapy to a waitlist control, or added on the energy therapy to a standard treatment to see whether it had additive effects, but not comparing it to a control group with a placebo added on. For the many reasons discussed in the previous chapter, having a randomized, placebo-controlled trial is crucial to being able to sort sense from nonsense in health care. This is especially true when examining those symptoms most likely to respond to placebos, such as pain and other subjective problems, which are, coincidentally, those for which energy therapies claim the greatest benefit.

In summary, energy therapies not only show no benefit in those few well-designed trials that have been conducted, but their basic underlying principles are not supported by evidence. There is no evidence to support the existence of the HEF, the manipulation of such a field, or the channeling of "energy" from one person into another.[10] Further, these concepts are in direct conflict with all that we know about how physics, chemistry, and biology work. As physicist Stenger (1999) wrote:

> Much of alternative medicine is based on claims that violate well-established scientific principles. Those that require the existence of a bioenergetic field, whether TT or acupuncture, should be asked to meet the same criteria as anyone else who claims a phenomenon, the existence of which goes beyond established science. They have an enormous burden of proof, and it is time that society laid it on their thin shoulders. (p. 9)

NATURAL PRODUCTS AND SUPPLEMENTS

If you've been to a pharmacy or grocery store in the past two decades, chances are you've wandered at some point into the health supplements aisle (or aisles, depending on the store). There you are surrounded by a visual cacophony of pills containing all manner of herbs, vitamins, and other substances ready to be taken to help you maintain optimal health. Or maybe you have a family member or friend who swears by a particular type of essential oil to treat spider bites, or help you sleep, or almost any other kind of problem you might have. What's the evidence base for such claims? Do we all need to be taking a good multivitamin and some supplements to live longer? In the last section of this chapter, we are going to broadly examine the "natural products" area of CAM.

The history of using plants, herbs, and other naturally occurring substances for medicinal purposes is likely as old as our species, especially given that other animals have been observed to eat particular plants to cure health problems.[11] We know that certain plants and herbs were used medicinally by at least the Paleolithic era, some 60,000 years ago, and written evidence shows the Sumerians, Egyptians, Chinese, and Indian cultures were using medicinal plants as far back as 5,000 years ago. Approximately 25% of modern pharmaceuticals are actually derived from plants in the first place,

---

[10] Technically speaking, there is always cannibalism, which certainly would transfer the chemical energy of one human into another via consumption, but that's not in any way what the energy therapies are talking about.

[11] For example, our closest relatives in the animal kingdom (chimpanzees and other apes) have been studied and found to eat certain plants when infected by intestinal parasites, to eat charcoal to counteract digestive problems, and more. This fascinating area of study (animals self-medicating via plant consumption) is called *zoopharmacognosy*.

with aspirin (coming from willow bark) and morphine (from the opium poppy) being the most well-known examples. Today, huge portions of the population of Asia, Africa, and Latin America use herbal medicines to meet primary health care needs (WHO, 2003). The largest survey to date in the United States found that almost 20% of Americans used herbal products, making it the most frequently used CAM modality (Barnes, Powell-Griner, McFann, & Nahin, 2004). Today, the use of herbal medicines is often referred to as phytotherapy.

Given the large amounts of public consumption and long history, one would guess that the herbs and plants used in phytotherapy have been extensively studied, especially given the approximately $15 billion spent on them each year in the United States alone (Millman, 2015). However, in a recent survey of the 1,000 most commonly used phytopharmaceuticals, only 156 of them had supportive clinical trials published, 724 had only basic science studies examining them, and 120 had absolutely no published data on their basic chemistry or application (Cravotto, Boffa, Genzini, & Garella, 2010). Furthermore, five of the 1,000 plants were actually toxic or allergenic, whereas only nine (or 0.009% of the total) of them were found to have "considerable evidence of therapeutic effect" (p. 11). In other words, well under 1% of these herbal medicines could be said to be EBT, with the remaining being non-EBT or PST.

The story behind why so many of the examined products of phytotherapy don't have evidence to support their use and why toxic ones can be sold in the United States is very interesting, and shows once again the intersection between politics and health. In 1994, two Senators, Republican Orrin Hatch of Utah and Democrat Tom Harkin of Iowa, authored and were able to pass the Dietary Supplement Health and Education Act (DSHEA). This Act contains regulations regarding the manufacture and sale of dietary supplements, which it defined as:

> A product (other than tobacco) intended to supplement the diet that bears or contains one of more of the following dietary ingredients: a vitamin, a mineral, an herb or other botanical, an amino acid, a dietary substance for use by man to supplement the diet by increasing the total dietary intake; or a concentrate, metabolite, constituent, extract, or combination of any ingredients.

The DSHEA further states that dietary supplements must be labeled as such, and that they are allowed to make one of three types of health claims:

1. Structure/function claims, which describe how a supplement impacts the body's structure or its function (e.g., "supports the immune system" or "arouses sexual desire" or "calcium builds strong bones")
2. General well-being claim (e.g., "helps you relax" or "relieves stress")

3. Benefit related to a classical nutrient deficiency disease claim (e.g., "Vitamin C prevents scurvy")

However, any supplement that makes such a claim also must include the disclaimer: "This statement has not been evaluated by the Food and Drug Administration. This product is not intended to diagnose, treat, cure, or prevent any disease." This is because the Food and Drug Administration (FDA) does *not* have to provide approval to a supplement manufacturer before the manufacturer can make and market these dietary supplements, unlike pharmaceutical manufacturers. In other words, the supplement manufacturers are not required to demonstrate that what they are selling is (a) effective or (b) safe before selling it the public. Instead, the FDA can only step in *after* a supplement has been found to be hazardous or dangerous.[12] The DSHEA is why, when less than 1% of the top selling herbal medicines have been found to be an EBT for particular conditions, they can all still be sold and marketed as being beneficial for a wide variety of problems.

If you are confused as to why two members of the Senate, supposedly charged with helping to protect the average consumer, would craft a bill that effectively deregulated an entire industry, then you don't know their backgrounds. Tom Harkin, shortly before the DSHEA was enacted, had pushed for the creation of the Office of Alternative Medicine (which later became the National Center for Complementary and Alternative Medicine) because he felt that taking bee pollen had cured him of his lifelong allergies. A staunch supporter of "natural" products and other types of CAM, Harkin helped increase funding for the agency he had created from $2 million during its 1992 establishment to $128 million in 2012. He even put language into the Patient Protection and Affordable Care Act of 2010 (known colloquially as Obamacare) to make sure that "licensed CAM providers" were recognized as official health care workers. Some of the top contributors to his campaigns over the past 20 years were the ACA, the Natural Products Association, and Herbalife International.

Hatch also counts Herbalife as one of his largest donors, and is the Senate's leading recipient of contributions from the dietary supplement industry as a whole. He happens to represent Utah, which is home to a huge number of major dietary supplement manufacturers, and has both family members and close friends who work as CAM practitioners or supplement industry lobbyists. Across his career, Hatch has repeatedly blocked attempts by other legislators to enact stricter regulation on the dietary supplement industry, as these would negatively impact his major campaign donors and a major industry of his home state. Clearly, Senators Harkin and Hatch were relying

[12] This is the equivalent of not only shutting the barn door after the horse has gotten out, but after the horse has gotten out and run over multiple people on its way to deposit a large sum of money at the bank.

on something other than scientific evidence to guide their support of "natural products" and herbal medicine when writing the DSHEA.

This deregulation has resulted in some fairly frightening effects. For instance, several investigations have found that many to most dietary and herbal supplements do not actually contain what their labels say they do. Newmaster, Grguric, Shanmughanandhan, Ramalingam, and Ragupathy (2013) used DNA barcoding to examine the contents of 44 different herbal products to see what, exactly, they contained. Their results showed that almost 60% of the products contained plant species that were not listed on the product's label, whereas a full third of tested products did not contain *any* of the plants listed on their bottle labels. This research was later replicated via an investigation conducted by New York state's attorney general, which found that 80% of the herbal supplements it tested (purchased from stores such as Wal-Mart, Target, Walgreens, and GNC) did not contain *any* genetic material from the plants listed on the label (O'Connor, 2015). Other research has shown high levels of products containing pharmaceuticals and/or heavy metals such as lead, mercury, and arsenic (Ernst, 2002). For an industry that is essentially supposed to be policing itself, the supplement manufacturers appear to be failing miserably.

These lax manufacturing standards, ineffective products, and potentially dangerous unlisted substances in dietary supplements lead many physicians and other health care leaders to urge great caution for anyone who plans to use an herbal product. It is also important to remember that herbs are, in fact, drugs. That is, they are substances containing multiple bioactive ingredients that can potentially be toxic, cause side effects, and even interact negatively with any other pharmaceuticals you are taking (Novella, 2013). This is one of the reasons most health care providers now ask about not just any medications you are currently taking, but also any nonprescribed supplements. Given how few have been shown to be EBTs, and that those few are not superior to conventional medications for the same conditions (only to placebos), it's not hard to see why almost all of phytotherapy falls under a non-EBT, pseudoscientific label. This is not to say that there are not valid medical uses of certain kinds of vitamins (e.g., taking iron to help with anemia), but these should be undertaken via sound medical guidance from a licensed health professional, not just randomly picked off a shelf because of their advertised benefits.

CONCLUSIONS

When it comes to physical health, people should be able to access providers and treatments that have been shown to be effective for specific problems. In the past few decades, more and more people in developed nations, who have access to modern medical care, have turned to complementary or alternative medical modalities to help maintain health or fix problematic health conditions. Unfortunately, as seen in this chapter, the overwhelming majority of such CAM therapies are either unproven or have reliably been shown to be ineffective for problems that they claim to be able to treat. Further, even for those CAM treatments that do have reliable evidence backing up their use (e.g., chiropractic for chronic, uncomplicated low back pain), they are no more effective than well-tested, efficacious "conventional" treatments. Considering the overblown and exaggerated claims, the lack of effectiveness demonstrated, and the potential danger inherent in some of these CAM treatments, the critical thinker should think long and hard before using them, especially if using them means one forgoes modern, conventional medical treatments.

QUESTIONS FOR REFLECTION

1. *Is there any scientific merit to subluxation, as used in chiropractic? If not, what reasons might explain the popularity of chiropractic treatment?*
2. *Interventions such as acupuncture have a very long history. Does the fact that acupuncture has been practiced for centuries have any bearing on its credibility?*
3. *Do you think that the public is deceived by practitioners of chiropractic being able to refer to themselves as "doctor?" Is this an ethical failing on the part of regulatory bodies or a case of "buyer beware" in action?*
4. *Given that live, attenuated viruses are used in vaccines like the MMR vaccine, is the homeopathic idea of "like cures like" really that unreasonable?*
5. *If natural products and vitamin supplements are largely useless, how can we explain their continued popularity? Have you ever used any? What for and why?*

REFERENCES

American Chiropractic Association. (2015). Key facts about the chiropractic profession. Retrieved from http://www.acatoday.org/pdf/Key-Facts-about-the-Chiropractic-Profession.pdf

American Chiropractic Association. (n.d.). History of chiropractic care. Retrieved from https://www.acatoday.org/level3_css.cfm?T1ID=13&T2ID=61&T3ID=149

Barnes, P. M., Powell-Griner, E., McFann, K., & Nahin, R. L. (2004). Complementary and alternative medicine use among adults: United States, 2002. *Advance Data From Vital and Health Statistics*, *343*, 1–19.

Biller, J., Sacco, R. L., Albuquerque, F. C., Demaerschalk, B. M., Fayad, P., Long, P. H., . . . Tirschwell, D. L. (2014). AHA/ASA scientific statement: Cervical arterial dissections and association with cervical manipulative therapy: A statement for healthcare professionals from the American Heart Association/American Stroke Association. *Stroke*, *45*, 3155–3174.

Bronfort, G., Nilsson, N., Haas, M., Evans, R., Goldsmith, C. H., Assendelft, W. J., & Bouter, L. M. (2004). Non-invasive physical treatments for chronic/recurrent headaches. *Cochrane Database of Systematic Reviews*, *3*, CD001878.

Campbell, J. B., Busse, J. W., & Injeyan, S. (2000). Chiropractors and vaccination: A historical perspective. *Pediatrics, 105*(4), 1–8.

Cherkin, D. C., Sherman, K. J., Avins, A. L., Erro, J. H., Ichikawa, L., Barlow, W. E., . . . Deyo, R. A. (2009). A randomized trial comparing acupuncture, simulated acupuncture, and usual care for chronic low back pain. *Archives of Internal Medicine, 169*(9), 858–866.

Colquhoun, D. (2008). Doctor who? Inappropriate use of titles by some alternative "medicine" practitioners. *New Zealand Journal of Medicine, 121*(1278), 6–10.

Colquhoun, D., & Novella, S. P. (2013). Acupuncture is theatrical placebo. *Anesthesia & Analgesia, 116*(6), 1360–1363.

Cravotto, G., Boffa, L., Genzini, L., & Garella, D. (2010). Phytotherapeutics: An evaluation of the potential of 1000 plants. *Journal of Clinical Pharmacy and Therapeutics, 35*, 11–48.

Dittrich, R., Rohsbach, D., Heidbreder, A., Heuschmann, P., Nassenstein, I., Bachmann, R., . . . Nabavi, D. G. (2007). Mild mechanical traumas are possible risk factors for cervical artery dissection. *Cerebrovascular Disease, 23*, 275–281.

Ernst, E. (2002). Toxic heavy metals and undeclared drugs in Asian herbal medicines. *Trends in Pharmacological Sciences, 23*(3), 136–139.

Ernst, E. (2010). Homeopathy: What does the "best" evidence tell us? *Medical Journal of Australia, 192*(8), 458–460.

Ernst, E., & Canter, P. H. (2006). A systematic review of systematic reviews of spinal manipulation. *Journal of the Royal Society of Medicine, 99*, 192–196.

Ernst, E., & Gilbey, A. (2010). Chiropractic claims in the English-speaking world. *New Zealand Medical Journal, 123*, 36–44.

Ernst, E., Lee, M. S., & Choi, T. Y. (2011). Acupuncture: Does it alleviate pain and are there serious risks? A review of reviews. *Pain, 152*, 755–764.

Forster, A. L. (1915). *Principles and practice of spinal adjustment; for the use of students and practitioners*. Chicago, IL: National School of Chiropractic.

Hammerschlag, R., Marx, B. L., & Aickin, M. (2014). Nontouch biofield therapy: A systematic review of human randomized controlled trials reporting use of only nonphysical contact treatment. *Journal of Alternative and Complementary Medicine, 20*(12): 881–892.

Huang, T., Shu, X., Huang, Y. S., & Cheuck, D. K. (2011). Complementary and miscellaneous interventions for nocturnal enuresis in children. *Cochrane Database of Systematic Reviews, 12*, CD005230.

International Center for Reiki Training. (n.d.). *What is Reiki?* Retrieved from http://reiki.org/FAQ/WhatIsReiki.html

Kaminskyj, A., Frazier, M., Johnstone, K., & Gleberzon, B. J. (2010). Chiropractic care for patients with asthma: A systematic review of the literature. *Journal of the Canadian Chiropractic Association, 54*(1), 24–32.

Karpouzis, F., Bonello, R., & Pollard, H. (2010). Chiropractic care for paediatric and adolescent attention-deficit/hyperactivity disorder: A systematic review. *Chiropractic & Osteopathy, 18*, 13.

Kohn, L. (2005). *Health and long life: The Chinese way*. Dunedin, FL: Three Pines Press.

Li, Z.-S. (2011). *The private life of Chairman Mao*. New York, NY: Random House.

Linde, K., Scholz, M., Ramirez, G., Clausius, N., Melchart, D., & Jonas, W. B. (1999). Impact of study quality on outcome in placebo controlled trials of homoeopathy. *Journal of Clinical Epidemiology, 52*, 631–636.

Long, P. H. (2013). *Chiropractic abuse: An insider's lament*. New York, NY: American Council on Science & Health.

Long, R., & Bernhardt, E. W. (1999). Perception of conventional sensory cues as an alternative to the postulated "Human Energy Field" of Therapeutic Touch. *Scientific Review of Alternative Medicine, 3*(2), 53–61.

Merepeza, A. (2014). Effects of spinal manipulation versus therapeutic exercise on adults with chronic low back pain: A literature review. *Journal of the Canadian Chiropractic Association, 58*(4), 456–466.

Millman, J. (2015, February 4). Americans are ignoring the science and spending billions on dietary supplements. *Washington Post*. Retrieved from http://www.washingtonpost.com/blogs/wonkblog/wp/2015/02/04/americans-are-ignoring-the-science-and-spending-billions-on-dietary-supplements

Mirtz, T. A., Morgan, L., Wyatt, L. H., & Greene, L. (2009). An epidemiological examination of the subluxation construct using Hill's criteria of causation. *Chiropractic & Osteopathy, 17*, 13.

National Board of Chiropractic Examiners. (2014). About chiropractic. Retrieved from http://www.nbce.org/about/about_chiropractic

National Center for Complementary and Integrative Health. (2009). *The use of complementary and alternative medicine in the United States: Cost data*. Retrieved from https://nccih.nih.gov/news/camstats/costs/costdatafs.htm

National Health and Medical Research Council. (2015). *NHMRC statement: Statement on homeopathy*. Retrieved from https://www.nhmrc.gov.au/_files_nhmrc/publications/attachments/cam02_nhmrc_statement_homeopathy.pdf

Newmaster, S. G., Grguric, M., Shanmughanandhan, D., Ramalingam, S., & Ragupathy, S. (2013). DNA barcoding detects contamination and substitution in North American herbal products. *BMC Medicine, 11*, 222.

Novella, S. (2009, June 24). Chiropractic—A brief overview, part 1 [blog post]. Retrieved from https://www.sciencebasedmedicine.org/chiropractic-a-brief-overview-part-i

Novella, S. (2013). Herbs are drugs. *Skeptical Inquirer*, 37(2).

O'Connor, A. (2015, March 30). GNC to strengthen supplement quality controls. *The New York Times*. Retrieved from http://well.blogs.nytimes.com/2015/03/30/gnc-to-strengthen-supplement-quality-controls/?_r=0

O'Mathúna, D. P., & Ashford, R. L. (2014). Therapeutic touch for healing acute wounds. *Cochrane Database of Systematic Reviews*, 7.

Palmer, B. J. (1909). *The philosophy of chiropractic, V*. Davenport, IA: Palmer School of Chiropractic.

Palmer, D. D. (1910). *The science, art and philosophy of chiropractic*. Portland, OR: Portland Printing House Company.

Park, R. L. (2008). *Superstition: Belief in the age of science*. Princeton, NJ: Princeton University Press.

Prasad, R. (2007). Homeopathy booming in India. *The Lancet, 370*(9600), 1679–1680.

Rosa, L., Rosa, E., Sarner, L., & Barrett, S. (1998). A close look at therapeutic touch. *Journal of the American Medical Association, 279*(13), 1005–1010.

Rothwell, D. M., Bondy, S. J., & Williams, I. (2001). Chiropractic manipulations and stroke: A population-based case-control study. *Stroke, 32,* 1054–1060.

Roy, V., Gupta, M., & Ghosh, R. K. (2015). Perception, attitude and usage of complementary and alternative medicine among doctors and patients in a tertiary care hospital in India. *Indian Journal of Pharmacology, 47*(2), 137–142.

Shang, A., Huwiler-Muntener, K., Nartey, L., Juni, P., Dorig, S., Sterne, J. A., . . . Egger, M. (2005). Are the clinical effects of homoeopathy placebo effects? Comparative study of placebo-controlled trials of homoeopathy and allopathy. *The Lancet, 366*(9487), 726–732.

Singh, S., & Ernst, E. (2008). *Trick or treatment? Alternative medicine on trial.* London, UK: Bantam Press.

Smith, W. S., Johnston, S. C., Skalabrin, E. J., Weater, M., Azari, P., Albers, G. W., & Gress, D. R. (2003). Spinal manipulative therapy is an independent risk factor for vertebral artery dissection. *Neurology, 60*(9), 1424–1428.

Spine Health. (2015). *Primary care providers.* Retrieved from http://www.spine-health.com/treatment/spine-specialists/primary-care-providers

Stenger, V. (1999). The physics of "Alternative Medicine" bioenergetic fields. *Scientific Review of Alternative Medicine, 3*(1), 1–10.

Therapeutic Touch International Association. (n.d.). *History of therapeutic touch.* Retrieved from http://therapeutic-touch.org/what-is-tt/history-of-tt

World Health Organization. (2003). *Traditional medicine.* Retrieved from http://www.who.int/mediacentre/factsheets/2003/fs134/en

Zhang, Y., Lao, L., Chen, H., & Ceballos, R. (2012). Acupuncture use among American adults: What acupuncture practitioners can learn from National Health Interview Survey 2007? *Evidence-Based Complementary and Alternative Medicine, 2012,* 1–8.

CHAPTER 12

PSEUDOSCIENCE IN MENTAL HEALTH

Whether from insurance providers, governmental agencies, professional organizations, or clients themselves, today's mental health practitioners (MHPs) are finding themselves under increasing pressure to justify not only what they do (e.g., providing psychotherapy or some other service), but also how they do it. In particular, just as in medicine, there has been a growing trend toward MHPs using evidence-based practice (EBP) in psychological treatment and assessment. As discussed extensively in Chapter 11, EBP is defined as "the conscientious, explicit, and judicious use of current best evidence in making decisions about the care of individual patients" (Sackett et al., 1996, p. 71). In the real world, this often translates into using therapies and assessment methods that have been demonstrated to be effective via valid and reliable clinical research.

EBP IN PSYCHOLOGY

Although interest in treatment efficacy has been present in applied psychology for decades, it wasn't until the mid-1990s that major movements toward EBP occurred in psychology as a whole. An American Psychological Association (APA) task force released a statement in 1995 detailing that the evidence for any psychological intervention should be based on two factors: efficacy (does research show it works?) and clinical utility (is it applicable in real-world settings?). Shortly thereafter, Diane Chambless and her associates (1996) thoroughly reviewed the literature and released the first list of empirically supported treatments (ESTs)—psychological treatments that had been found to have high levels of efficacy for specific disorders, generally comparable to or exceeding the effects of medication. Included were methods such as cognitive behavioral therapy (CBT) for panic disorder, exposure with response prevention (EX/RP) (a type of behavioral therapy) for obsessive-compulsive disorder, and interpersonal therapy for major depression. The response to this publication was varied, with some praising the effort as a bold move to raise public awareness of the efficacy of psychological treatment. Others, though, complained about the lack of focus on common, nonspecific therapeutic factors and the emphasis on treatment manuals and short-term therapy. Despite this criticism, the push to increase the use of EBP within the psychological community continued to grow.

In 2006, the APA's Presidential Task Force on EBP furthered its commitment to EBP in the field of psychology via integration of applied and basic research. In a resulting article (APA, 2006), EBP was defined as "the integration of the best available research with clinical expertise in the context of patient characteristics, culture, and preferences" and the purpose of EBP was "to promote effective psychological practice and enhance public health by applying empirically supported principles of psychological assessment, case formulation, therapeutic relationship, and intervention" (p. 273). There have been hundreds of clinical trials investigating the efficacy of particular forms of EBP, such as cognitive, behavioral, and interpersonal therapies, with more emerging all the time. In addition to clinical efficacy, research has shown time and time again that EBP is safe and effective for a wide range of ages and problems, is more enduring in symptom impact than medications, and pays for itself via medical cost offset and increased productivity, providing a huge support base for the clinical utility of such treatments (Lambert & Ogles, 2004).

Unfortunately, a large number of MHPs have little to no training in how to utilize evidence-based psychotherapies and assessments. Especially in low income and rural areas, there are few doctoral-level practitioners available, which means that the majority of services are being delivered via master's-level MHPs. Traditionally, clinical psychology doctoral programs and postdoctoral positions have been the primary place where practitioners received extensive training in CBT, interpersonal, and other evidence-based therapies. Although there are a growing number of master's-level programs that emphasize EBP, they are in the minority. This difference in training level is critical, as there are approximately 665,500 master's-level counselors and 642,000 social workers in the United States, compared with 152,000 doctoral-level clinical and counseling psychologists (U.S. Department of Labor, 2010). This means that a majority of MHPs are not well trained in EBP, which opens the door to providers plying pseudoscientific, non–evidence-based treatments (non-EBTs) and assessments for mental health problems. Likewise, many people seeking psychological services do not know the difference between EBP and non-EBP, making them vulnerable to receiving services that are not effective. The purpose of this chapter is to shine the light of critical inquiry onto some of the most commonly seen non-EBP in psychology and counseling, from treatments for specific disorders, such as autism, posttraumatic stress disorder (PTSD), and substance abuse, to psychological assessment measures.

AUTISM SPECTRUM DISORDER

Autism spectrum disorder (ASD) is characterized by impairments in social interaction and communication (verbal and nonverbal), frequently accompanied by repetitive self-soothing behavior. Individuals with ASD experience difficulty connecting to others in social situations, have difficulty experiencing empathy, and often have issues with impulsive behavior and self-regulation, especially when confronted with novel stimuli. In the recent past, the disorder was seen as relatively rare, with only the most extreme cases of nonverbal autism attended to by the health care community. Over the course of the past 25 years, though, a far greater number of individuals are being diagnosed with ASD (caused in large part by the expansion of the official diagnostic criteria and increased screening and awareness). This has resulted in increased interest in treating the disorder through both medical and psychological methods, but has also resulted in an explosion of pseudoscientific and non-EBTs being offered to the public.

Autism and Vaccines

As in many areas of psychology, there is a large gap between what we actually know about ASD via research and what a number of people believe. Probably the most dangerous myth surrounding autism is that childhood vaccinations can somehow cause ASD. This thoroughly discredited idea started largely on the basis of a single, poorly conducted study published in 1998 by Andrew Wakefield, a medical doctor in the United Kingdom. Wakefield, along with 12 coauthors, claimed that in a study of 12 children they observed a combination of gastrointestinal disease and developmental regression (both of which are often seen in children with ASD, but neither of which is a diagnostic indicator) that occurred within 2 weeks of receiving a measles, mumps, and rubella (MMR) vaccine. It is important to note that there was no causal explanation provided, even a hypothetical one, in the original work. For Wakefield, this was enough to call a press conference and to speculate to the media about the safety of the widely used MMR vaccine. Since that time, controversy has reigned within popular culture regarding the potential of vaccines to cause autism, among other disorders and ailments.

In 2004, when media scrutiny of the study first began to build, 10 of Wakefield's coauthors retracted their names from the interpretation in the work, and distanced themselves from any conclusions that they had initially approved. The scrutiny and disavowals largely came in response to a news story published in the *Sunday Times*, in which parents of the original 12 children in the study were found to have been recruited by a lawyer preparing a massive lawsuit against manufacturers of the MMR vaccine, and that this same lawyer had given over £400,000 to Andrew Wakefield. To think of this as anything other than a quid pro quo deal would be naïve, at best. Without speculating on any prior arrangements, it is evident that Wakefield's clear conflict of interest prevented him from conducting the study in an unbiased and objective manner.

Within months, additional evidence was levied against Wakefield, namely, that he had applied for a patent prior to publishing this study in *The Lancet*. This patent was for a single-jab measles vaccine that would allegedly prevent autism "caused" by the currently used MMR vaccine. So not only was Wakefield being paid to conduct the research by a lawyer preparing a lawsuit against vaccine manufacturers, using participants recruited by this same attorney, but he had also applied for a patent for the very vaccine that would be needed if his research was found to be accurate. Dr. Wakefield was severely compromised, something that the General Medical Council (the British medical regulatory board) took note of, conducting a "fitness to practice" trial from July 2007 to May 2010. In addition to the conflict-of-interest charges that were brought against him regarding his relationship with the lawyer, facts regarding the treatment of his original subjects came to light that cast further doubt on Wakefield's ability to practice medicine. The children had been subjected to a number of painful invasive tests without any medical cause, such as spinal taps, colonoscopies, and colon biopsies *without* first obtaining the approval of an ethics board. The board ruled against Wakefield on all charges (and several unlisted ones that seem relatively minor in comparison to the vaccine-panic issue) and removed him from the UK medical register, rendering him unable to practice medicine in the United Kingdom.

These investigations and facts led *The Lancet*, the scientific journal that originally published his study, to completely retract his work in 2010, stating that his conclusions were "fatally flawed." Despite this thorough discrediting, the damage had been done. Numerous celebrities, such as Jenny McCarthy, Jim Carrey, Mayim Bialik, Robert F. Kennedy, Jr., and Bill Maher, took up the antivaccination cause, resulting in significant drops in vaccine rates in the United Kingdom and United States. In 1996, 92% of all individuals 24-months old in the United Kingdom were vaccinated for MMR. By 2002 it was 84%, with some parts of London dropping all the way to 61% in late 2003. Although vaccination rates have rebounded since Wakefield's fall, measles is now endemic in the United Kingdom, meaning it survives within the population. Thousands of people have gotten sick, and dozens have died in the outbreaks that have occurred. In the United States, measles was declared eliminated in 2000, but by 2013 there were major outbreaks in three states as a result of nonvaccinated children, with increasing outbreaks in the years 2014 and 2015 in dozens of states. The first death as a result of measles in the United States in over a decade occurred in 2015 and can be directly attributed to the antivaccination movement.

The worst part about the thousands of people who have been needlessly infected or killed by measles (not to mention the whooping cough and other preventable disease outbreaks that have occurred recently) is that they all sprang from a single shoddy study, while other, better controlled studies that showed no link between vaccines and ASD (and other health problems) were ignored. In fact, the largest ever study to examine the potential link between MMR vaccines and ASD used the health records of almost 96,000 children (compare that number to the 12 in Wakefield's discredited study). The researchers found that not only did the vaccine not increase the risk of a child developing ASD, but that this held true even when a child had an older sibling who was already diagnosed with ASD (Jain et al., 2015). This reinforced findings from dozens of smaller studies over the past two decades. This replication stands in sharp contrast to those studies that tried to confirm Wakefield's results and found nothing similar to what he had found.

Evidence Based Versus Pseudoscientific Treatments for Autism

Given the rise in diagnoses of ASDs, there has been a corresponding increase in the numbers of parents seeking effective treatment for their child's condition. In terms of broad and replicated findings, research has repeatedly shown that early, intensive treatments using techniques of applied behavioral analysis (ABA) can help make substantial gains in measured intelligence, academic skills, language abilities, and adaptive functioning for children with ASD (Weitlauf et al., 2014). Despite this, ABA tends to be underutilized because of a lack of qualified practitioners and knowledge about the treatment on the part of parents and physicians. This has led to a proliferation of other therapeutic modalities to help those with ASD, despite not having the demonstrated effectiveness of ABA.

Facilitated Communication

With communication so vastly impaired in people with autism, it comes as little surprise that many parents desperately seek a treatment to address that issue. In the early 1990s, many therapists and parents thought they had found an amazing therapy to help their severely impaired children talk to them, called facilitated communication (FC). The idea behind FC (sometimes called "supported typing") is that persons with autism or other severe developmental problems are not able to communicate effectively because of motor coordination issues, rather than some type of brain-based neurological problem (American Speech-Language-Hearing Association, 1994). Supporters of FC contend that these individuals need specialized physical, communicative, and emotional support in order to communicate effectively.

To accomplish this, a trained facilitator supports the hand or wrist of the communicator (the person with ASD) while it hovers over a keyboard, allowing the person with ASD to overcome these motor issues and point to letters or words. This support can vary from just holding a single finger to holding the wrist, elbow, or shoulder (or a combination of those). The communicative support involved may be things like rephrasing questions to make them simpler or more predictable. The emotional support involves being encouraging and patient while assuming the communicator actually can communicate with the right help. For many parents who took their children for FC sessions, the results were astounding: children who had never spoken aloud were suddenly typing out phrases such as "I love you," complex sentences, and even poetry. Parents were rejoicing at finally being able to interact on a more meaningful level with their children and facilitators could bask in knowing that they were making a huge impact on people's lives.

FC provides a model case study for the use of critical-thinking skills when it comes to health claims. First, the idea that children could not communicate because of motor and coordination difficulties rather than speech and language problems is an extraordinary claim, which flies in the face of all available evidence about what causes communication problems in those with ASD. For instance, although a number of children with ASD have less developed motor skills than their same-age peers, many of these children who needed "help" to type would have had no problem with other relatively complex motor tasks, such as feeding themselves. Next, we can evaluate rival hypotheses for why FC "works." One, held up by users of FC, is that the physical and emotional support offered allows the child or adult with ASD to communicate. A second (and more parsimonious) hypothesis is that the facilitators are actually the ones who are doing the communicating. In this case, relatively simple experiments can be performed to help determine which of these hypotheses is more plausible.

One way to rule out the influence of the facilitator is to make the facilitator blind to what the response should be. For example, one could put headphones on both the facilitator and communicator and then play questions and ask them to answer. But you could also blind the test by making half of the questions the same (e.g., both hear "What color is the sky?") and make the other half different questions (e.g., "What color is the sky?" to the facilitator and "What color is grass?" to the communicator). You would then take and compare the answers to see if what is typed out answers the question posed to the facilitator or communicator. In another simple blinded test you can substitute auditory questions for visual pictures, again to see if what is typed out is congruent with what the communicator or the facilitator is shown. It turns out that when studies have used these and other control methods to rule out facilitator influence, they find time and again (consistent with the solid replication needed in science) that the answers typed out are inevitably those of the facilitator (Romancyzk, Turner, Sevlever, & Gillis, 2015).

These studies have led to FC being labeled pseudoscientific and harmful by organizations from the American Academy of Pediatrics to the APA. In a 1994 report, the American Speech–Language–Hearing Association said "Results of experimental investigations consistently fail to support the validity of FC, and provide repeated examples of facilitators unknowingly influencing the content of messages that they believed were conveyed by the FC user." Unfortunately, this has not stopped people from using it and claiming massive success, despite the warnings of professional associations and a lack of an evidence base.

On the parents' side, this may be in part because of the confusion between FC and what is called "augmented communication" (AC). In AC, the communicator always has *direct* voluntary control of the means of communication, such as having a tablet with symbols or words on it that he or she points to or using small hand/head movements to shift a cursor around a screen (not unlike what Stephen Hawking uses). The key difference is that the communicator is always in control with AC, which has been demonstrated repeatedly in well-controlled research. In FC, it is actually the facilitator who is in control, which is what our blinded trials reveal.

Sensory Integration Therapy

Sensory integration therapy (SIT) is a treatment often employed by occupational therapists to help children with autism. It is intended to address higher level cognitive impairments in functioning by fixing what are called "sensory processing problems." It reportedly does this by engaging in activities to stimulate the vestibular (movement), proprioceptive (motor), and tactile (touch) senses that will balance out these problems. Used since the 1970s, proponents of SIT often say that it is grounded in contemporary neuroscience's understanding of how our brains process information. However, as with FC and many other non-EBP ideas, SIT's core theories have no empirical support. Not only are the core ideas out of sync with our understanding of how sensory perception is related to higher cognitive functioning, the idea of a disorder of sensory processing is not accepted by the American Medical Association or the APA. The lack of evidence base is further compounded by positive results only coming from poorly designed and poorly controlled studies, with proper studies finding "no objective, scientific evidence to suggest that SIT is more effective than alternative treatments or even *no treatment at all*" (Polenick & Flora, 2012, p. 35).

Biological Treatments

The final two alternative treatments we discuss for ASD are not only ineffective, but also potentially dangerous: chelation therapy and megadoses of vitamins. Chelation therapy is a procedure for removing toxic heavy metals such as mercury, arsenic, and lead from one's body. Typically used when someone accidentally ingests heavy metals, the process involves administration of a chelation agent (what kind depends on the type of metal that needs to be removed) that binds to the toxic metal and is then excreted by the body. This can be a very dangerous treatment, with side effects ranging from kidney damage to allergic, congestive heart failure, liver impairment, and even death via cardiac arrest (Atwood, Woeckner, Baratz, & Sampson, 2008).

Despite only being approved to treat acute heavy metal poisoning, some people have begun marketing it as a treatment for ASD, on the false assumption that autism symptoms are caused by mercury exposure from childhood vaccinations[1] and dental fillings. The evidence is clear that there is no actual link between vaccines and autism symptoms, but that has not stopped hundreds of practitioners from promoting chelation therapy as an ASD treatment. The tragic aspect here is that, unlike SIT, chelation can be deadly, and sadly has been. In 2005, a 5-year-old British boy named Abubakar Tariq Nadama died while receiving chelation in a Pennsylvania physician's office. Although the death was determined to be the result of a drug mix-up, Tariq had been brought to the United States by his mother specifically to get chelation therapy to help cure his autism (Kane, 2006). Despite the lack of a sound underlying theory for its use, no evidence that it can alleviate the symptoms of ASD, and the inherent danger in the treatment, chelation therapy is still heavily promoted and advertised as an "alternative" way to treat autism by many practitioners and websites.

Many children with ASD display very picky eating habits, and as a result over half of parents give them vitamin supplements either because they do not think their children are getting a well-rounded diet or because their pediatricians recommend it. Others give megadoses of vitamins like B_6 in the belief that it will help improve behavior and cognitive skills. Research has found that a significant number of children with ASD are deficient in certain micronutrients (particularly vitamin D, calcium, and potassium). However, supplement use often failed to correct those deficiencies and instead led to excesses of others, such as vitamins A and C, copper, zinc, and manganese (Stewart et al., 2015). Well-designed research trials that have examined B vitamin supplementation have not found any significant behavioral

[1] Childhood vaccinations had contained thimerosal, a mercury-based preservative, since the 1930s. This practice was stopped in 1999 purely as a precautionary measure in response to a Food & Drug Administration review, which means that even if it had caused autism (which it didn't) that would have ended over 15 years ago. Also, the symptoms of mercury poisoning are well known, and are significantly different from those symptoms seen in ASD.

or cognitive impacts from taking the supplements (Romancyzk et al., 2015). Given the known negative side effects and toxicity from excess vitamins (including A, B_6, and C), providing these children with vitamins should only be done after careful consultation with a physician and should only target existing, identified deficiencies to avoid excess intake.

In conclusion, there is no cure or "magic bullet" for ASD, despite the claims of those who are willing to sell treatments shown to be ineffective and potentially dangerous. There is an EBT called ABA that can improve behavior, communication, and intelligence, but it relies on early, intensive behavioral interventions based on the known and well-studied principles of learning. Parents searching for help for their children with ASD need to carefully wade through the morass of non–evidence-based pseudoscience in order to gain the most benefits possible.

TRAUMA-FOCUSED THERAPIES

Over the last 25 years, as news media outlets have proliferated and information takes less and less time to travel around the world, people have been bombarded with images and reports of traumatic events taking place all around the globe. Although most of our minds immediately turn to man-made traumas, such as the Columbine massacre, the "Dark Knight" shooting in Colorado, or the terrorist attacks of 9/11, natural disasters such as earthquakes, tornadoes, and hurricanes can cause reactions that are equally impairing for adults and children (Lack, 2008). In the media, however, the focus is almost exclusively on immediate, short-term reactions, with little reporting on the long-term impact of a disaster. Unfortunately, this is also the case in most interventions: a strong response immediately postdisaster is followed by a lack of preparation for or an inability to deal with the potential psychological, emotional, and behavioral disturbances seen in a number of children and adults after a traumatic event (Jaycox et al., 2007).

The most common difficulty experienced by people after a disaster or trauma is some form of anxiety, with posttraumatic stress symptoms being the most common type (Bland et al., 2005). Related impairments in social and academic functioning, as well as other mental health impairments such as depression and substance abuse are also frequently seen. In children, trauma exposure is related to cognitive impairments, lowered grades, increased school absences, and lowered graduation rates. In addition, posttraumatic stress symptoms include many school- and work-impairing difficulties, such as problems concentrating, sleep disturbance, and disorganized behavior (American Psychiatric Association, 2013).

Although there is no shortage of well-intentioned practitioners willing to provide services for those struggling with ongoing posttraumatic stress symptoms, it is crucial to deliver and receive evidence-based rather than pseudoscientific interventions in the aftermath of trauma. Despite the enormously strong evidence base supporting the use of cognitive and behavioral techniques for treating posttraumatic stress (Amaya-Jackson et al., 2003), there are nonetheless many proponents of other, non–evidence-based therapies that either have no research support or evidence against their use. Four of the most widespread pseudoscientific treatments for trauma-related problems are critical incident stress management (CISM), eye movement desensitization and retraining (EMDR), emotional freedom technique (EFT), and thought field therapy (TFT).

Critical Incident Stress Management

Widely known as "psychological debriefing," CISM (Mitchell & Everly, 1998) developed in the early 1980s and, unlike the other interventions described in this section, was focused on the prevention of PTSD symptoms, rather than their treatment. CISM is based on assumptions that (a) trauma exposure alone is enough to cause a person to experience long-term psychological difficulties and (b) early interventions can prevent such problems from developing. However, carefully controlled studies have found that the vast majority of people recover without any interventions after a trauma, showing little to no distress at 3 months posttrauma (e.g., Ehlers, Mayou, & Bryant, 1998). In addition, numerous scientific studies have found that receiving CISM appears to actually increase the chance someone will develop PTSD symptoms (for a review, see McNally, Bryant, & Ehlers [2003]). All evidence supporting the use of CISM is based on anecdotal reports and is primarily published by the originator of the method, J. T. Mitchell. Based on the, at best, inert effects and, at worst, harmful impact of CISM, numerous organizations, including the World Health Organization and British National Health Service, have actively implemented policies against its use. In short, "Although psychological debriefing is widely used throughout the world to prevent PTSD, there is no convincing evidence that it does so" (McNally et al., 2003, p. 72).

Eye Movement Desensitization and Reprocessing

Eye movement desensitization and reprocessing (EMDR) is one of the most heavily promoted and commercialized pseudoscientific psychological treatments of the last 30 years (Lohr, Gist, Deacon, Devilly, & Varker, 2015). Developed in the late 1980s by Francine Shapiro, EMDR is a very structured intervention that huge numbers of clinicians have reportedly been trained to use. It starts with taking a detailed history of one's traumatic event (or events) and the symptoms a person is currently experiencing. From there, the therapist instructs the client to vividly imagine the trauma, including how he or she physically felt. While holding those memories and feelings, the therapist then introduces "bilateral stimulation" (originally moving one's eyes from side to side, but now it's been expanded to physical tapping on both sides of the body or even having tones played first in one ear, then another). Doing so reportedly "unlocks" the brain through this "dual attention" procedure, resulting in a reduction of PTSD symptoms.

There have been a number of investigations and well-controlled research into the efficacy of EMDR. Large-scale analyses of these trials show that EMDR appears to work about as well as well-established cognitive behavioral therapies for PTSD that focus on the use of EX/RP. Based on this, the American Psychiatric Association has stated, "EMDR appears to be effective in ameliorating symptoms of both acute and chronic PTSD" (Work Group on ASD and PTSD, 2004, p. 59). This statement is advertised prominently on the EMDR Institute's homepage, but the page omits the other, less favorable conclusions of the American Psychiatric Association's review of research on EMDR, which states that "Despite the demonstrable efficacy of EMDR, these studies call into question EMDR's theoretical rationale" (Work Group on ASD and PTSD, 2004, p. 59).

Specifically, many researchers and theorists see EMDR as a prime example of a "purple hat therapy" (Rosen & Davison, 2003). Purple hat treatments take something that is known to work for a particular problem, such as EX/RP for PTSD symptoms, and then add on another element, such as making the client wear a purple hat during the treatment. Then, when the treatment works, they attribute the success not to the already known active change agent, but to the magical purple hat. In EMDR, the active ingredients causing change are use of CBT techniques, such as EX/RP, and not the use of bilateral eye or body movements, which are considered a "key" component of the treatment package (Shapiro, 1995). Dismantling research, which breaks treatments down into their component parts to see which of those pieces are actually causing change, has shown that removing the purple hat of bilateral movements makes the treatment no less effective at treating PTSD (Spates et al., 2009). In short, EMDR works, but it doesn't work because of why it purports to work. Instead, it works because of the EX/RP component. This is what led one author to write that "What is effective in EMDR is not new, and what is new is not effective" (McNally, 1999, p. 619).

"Energy" Trauma Therapies

TFT (sometimes called the Callahan Techniques) is a treatment based in part on traditional Chinese medicine, relying on the idea that invisible energy fields, or "thought fields," surround the body (Guadino & Herbert, 2000). By physically tapping on places in the body where these fields intersect one can supposedly modify these thought fields and cause a decrease in negative emotions, similar to how acupuncture supposedly relieves physical pain (Callahan & Callahan, 1996). The main TFT website touts it as highly effective, saying, "TFT often works when nothing else will. . . . It has been used for weight loss, stop smoking [*sic*], phobias, trauma relief, love pain, and much, much more" (rogercallahan.com, n.d.). It goes on to say that "When applied to problems TFT addresses their fundamental causes, providing information in the form of a healing code, balancing the body's energy system and allowing you to eliminate most negative emotions within minutes and promote the body's own healing ability."

But as we saw with acupuncture in the previous chapter, there is no scientific support for this theory of "energy" flowing through the body. Further, there is no sound outcome research supporting the efficacy of TFT for treating any emotional disorder, despite the claims of TFT's proponents (Guadino & Herbert, 2000). This goes doubly so for EFT (Craig, 1997), which evolved from TFT and presents itself as even more comprehensive, even with no research to support its claims (Waite & Holder, 2003). With the inability to falsify their claims, a reliance on anecdotal evidence, and claims of miraculous success (such as those on the websites of the World Center for EFT; http://www.emofree.com.au/home/what-is-eft)—"The surprising natural healing aid you can use for almost everything"—and Callahan Technique's (http://www.rogercallahan.com/home.php) "balancing the body's energy system and allowing you to eliminate most negative emotions within minutes"), there is no reason to believe these therapies are anything other than pseudoscience.

SUBSTANCE USE TREATMENT

The excessive use of substances, whether alcohol, nicotine, prescription painkillers, methamphetamines, or any other drug, gives rise to a booming industry in the United States and abroad. The National Institute on Drug Abuse reports that in the United States over $600 billion is spent annually in drug-related crime, work loss, and health care costs. Alcohol and tobacco (at 85% and 81% lifetime use rates, respectively) are used more frequently than any other drugs, with annual costs related to their use at approximately $428 billion dollars, of which $126 billion are health-related costs, such as emergency room visits, treatment programs, and other various associated costs (National Institute on Drug Abuse [NIDA], 2015). Marijuana (39.1%) and cocaine (14.4%) are the most frequently used illicit drugs, whereas nonmedical use of psychotherapeutic drugs (19.1%) and pain relievers (15.4%) have high usage rates as well. Global estimates are that alcohol abuse results in around 2.5 million deaths annually, whereas tobacco use is responsible for over 5 million deaths each year (World Health Organization, 2014). Given these huge costs, it only makes sense to apply the most effective EBTs to these problems, rather than waste time and money on something either shown not to work or of unknown efficacy. Unfortunately, though, the most widespread and familiar substance use treatment is not actually an EBT.

Although Alcoholics Anonymous (AA) groups are one of the oldest and most popular methods of rehabilitation for addiction problems, having spawned a legion of similar "anonymous" groups (e.g., Narcotics Anonymous, Gamblers Anonymous). AA and its sister groups are based around a fellowship of individuals helping each other get through their substance use problems. Their self-help 12-step recovery method has become deeply ingrained in society as the key method for treating addiction since the program's inception in 1935. The effectiveness of this program is so widely accepted that many courts will mandate that people with alcohol and other substance abuse disorders attend AA meetings. Unfortunately, these 12-step programs are founded on shaky science and anecdotal evidence, not a solid evidence base.

The founding principles of AA are religious in nature, derived from the teachings of the Oxford Group, a Christian sect (many would use the word *cult* in this case) started by the evangelical minster Frank Buchman in the 1920s. The Oxford Group's main tenets are at the core of what William Griffith Wilson (Bill W.) and Robert Holbrook Smith (Dr. Bob) later developed into the 12 steps of AA. These included sharing one's "sins and temptations," giving one's life over to God, attempting restitution to people you had wronged, and looking for guidance from God. In addition, one was directed to examine his or her own life closely and then proselytize to others about the Oxford Group.

The 12 steps of AA continued on this religious path, with 7 of the 12 steps referencing God or "Him." So although AA today often claims to be agnostic, saying it does not require religious faith to "work," this is in direct conflict with its founding principles, which frequently refer to God (or a "higher power") and one's relationship with God as central to recovery. Multiple steps tell the substance abuser to give himself over to God in order to be healed. There is no evidence-based scientific medical or psychological practice in the 21st century that says one needs to give oneself up to God for recovery to work. Yet, regularly, this is what people with substance use disorders are told to do if they want to get better. Recovery based on religious principles and prayer may be important for some, but it could simultaneously prove to be a detriment to a nonreligious person seeking support during recovery. In fact, one survey showed that 66% of former AA members disliked the religious aspects, and over half found it to be the least helpful aspect of the program (Connors & Dermen, 1996). Moreover, when research examined what parts of AA contribute the most effects, it showed that the religious and spiritual aspects of AA are not what cause change for the minority of people who do change. Instead, the reason why AA works for anyone is the development of a healthier social network (e.g., hanging out with fewer heavy drinkers), increased self-efficacy (confidence in one's ability to exert control over themselves and their environment), and increased coping skills (Kelly, Magill, & Stout, 2009). These features are not unique to AA, and are consistently found in effective mutual self-help groups such as SMART Recovery, Secular Organizations for Sobriety, or Moderation Management.

Another, perhaps more damning flaw of these programs is that they do not seem to work for most people, based on all available data. According to the AA triennial survey, 76% of their membership has been sober for over a year (AA, 2014). However, research on AA from people outside of the organization shows a very different result, with success rates of between 5% and 20% (Ferri, Amato, & Davoli, 2006). For a hard-to-treat behavioral problem, this still sounds good, until you realize that spontaneous recovery (getting better with no intervention) happens in 26.2% of individuals with an alcohol abuse problem during any given year (Walters, 2000). This type of regression to the mean is rarely accounted for in the pro-AA literature but is crucial when examining whether a treatment is truly effective.

But even with such low success rates, 40% of AA members were referred by a health care professional and 12% were court-mandated to attend. This illustrates, once again, how many MHPs, health care providers, and even our legal system fall prey to believing something works well when it does not. A final problem with the AA/12-step model is that it demands complete and total abstinence from drinking (or using any other drug) from its participants. This is in sharp contrast to decades of available research showing that substantial portions of once-problematic drinkers are actually able to consume alcohol in moderation with no resulting functional impairment (MacKillop & Gray, 2014).

One should note, though, that the takeaway from this section of the chapter should not be that "AA doesn't work." Although there are no clinical trials demonstrating that it is effective, the research does appear to show that AA-type groups work fairly well for a particular group of people—those who strongly identify with the spiritual concepts inherent in the program. The aspects of AA that appear to be helpful in reducing problematic drinking, such as improving coping skills; developing a supportive, nondrinking social network; and improving one's self-efficacy, are common to many mutual self-help programs. Moreover, these types of mutual self-help groups seem to operate most effectively when paired with EBTs for substance abuse such as CBT, motivational interviewing, relapse prevention, and even some types of medications (or combinations of these; MacKillop & Gray, 2015). So the takeaway should be: AA can work for certain people (not everyone), but not because of something unique to it, and it should be combined with individualized, evidence-based treatment.

PROJECTIVE TESTING

The assessment of an individual's psychological state—thoughts, feelings, and behaviors—is an enormous part of a mental health clinician's life. It helps guide treatment planning, decisions in court cases, and more. But, not all psychological assessments are created equal. Types of measures used to assess for personality characteristics and psychopathology are often divided into two categories: objective and projective (Weiner & Greene, 2008). Objective tests make *direct* inferences about a person's psychological state based on his or her self-report (or in some cases, report from significant others such as parents) to very clear questions. Projective tests, in which instructions or stimuli are more ambiguous and less structured, make *indirect* inferences about a person's psychological state. The term "projective" itself comes from Frank (1939), who thought that using ambiguous stimuli would allow a person to project his or her "private world" onto such stimuli, and as such "interpret the material and react affectively to it" (p. 403). In this section, we discuss the origins of three common types of projective tests, focusing on their theoretical underpinnings[2] and the scientific evidence for such theories, and the support for their use in clinical settings.

Reliability and Validity

But before we move into that, it's important to define two terms that are integral to determining whether any type of diagnostic tool (medical or psychological) is useful. First is *reliability*, which refers to the consistency or stability of results. There are many different ways that reliability is determined. One example would be whether a test gives you the same result when administered to the same person at a different time (test–retest reliability, measuring temporal stability). Another is to have two people give the same test to the same individual, to see whether their interpretations of the tests match up (inter-rater reliability). The second term is *validity*, which refers to whether or not the test measures what it claims to measure. For example, if we were to give you a test that was supposed to measure the level of sodium in your blood, but instead measured the amount of potassium, it would not be a valid test.

A test must be reliable to be considered valid, but reliability alone does not determine validity. In other words, a test can be reliable without being valid, but it cannot be valid without being reliable. If you are shooting arrows at a target, you try to get as close to the middle (the bull's-eye) as you can. Similarly, in diagnostic work, we try to get as accurate a diagnosis as possible. Using the targets in Figure 12.1, we can see different types of tests illustrated. On the left we can see a test that has demonstrated reliability (meaning, the arrows all cluster together in roughly the same place), but not validity (they are far from the bull's-eye). In the middle, we have a test

[2] Many persons who are only superficially familiar with the development of the various measures to be discussed in the following (the Rorschach Inkblot Method, the Thematic Apperception Test, and figure drawings) have the idea that the usage and interpretation of these measures are all based on Sigmund Freud's theories of personality and psychoanalysis. This, however, is far from the truth. In fact, each measure described has its own unique development, sometimes directly related to Freudian theories, sometimes influenced by them, and sometimes largely independent of them.

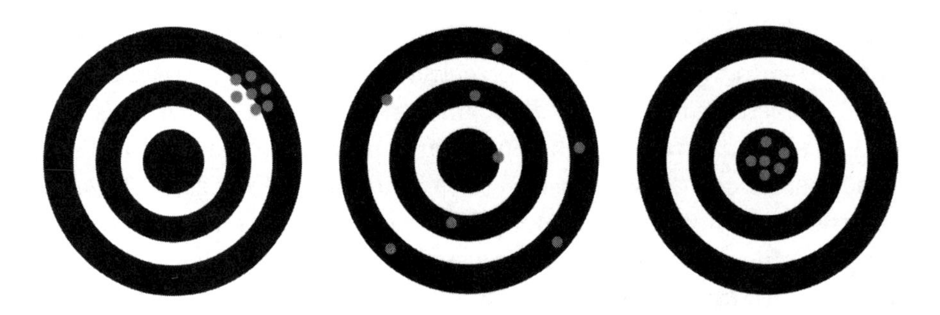

FIGURE 12.1 Demonstrations of validity and reliability.

with neither reliability (the arrows are all over the target) nor validity (only one is in the bull's-eye). On the right we see a test that demonstrates both reliability (the arrows are all close together) and validity (they are all within the bull's-eye). It's only when you have a reliable and valid test that you get accurate diagnoses.

Rorschach Inkblot Method

To gain an understanding of the strength of beliefs for and against the use of our first test, the Rorschach (also called the Rorschach Inkblot Method), which has been described as being "the most cherished and the most reviled of all psychological assessment tools" (Hunsley & Bailey, 1999, p. 266). It is often listed as one of the most commonly used psychological measures by clinical and school psychologists (Hojnoski, Morrison, Brown, & Matthews, 2006), although anecdotal experience suggests a decline in use across the past decade. The Rorschach also holds a grip on the public imagination, as evidenced by the use of similar inkblots in media from comic books ("Watchmen" by Alan Moore and Dave Gibbons) to music videos ("Crazy" by Gnarls Barkley).

Hermann Rorschach's development of the test that would bear his name provides an interesting story. Rorschach created his blots and developed their usage in 1918, apparently inspired by a popular parlor game called *Klecksographie* (roughly "Blotto" in English), in which one would drip ink onto a piece of paper, fold it in half, and then compete to give the most numerous or interesting answers (Exner, 2003). Rorschach's inkblots were not what one would have seen in a game of Blotto, though, as he painstakingly constructed them using ink and watercolors, rather than relying purely on chance or random drips and patterns (Morganthaler, 1954). To gain an idea of what these inblots look like, we constructed Figure 12.2 using the same methods as Rorschach. Based on his only major work (he died at age 37, only 9 months after publication of it), Rorschach was particularly concerned with two factors in a person's response to the blots:

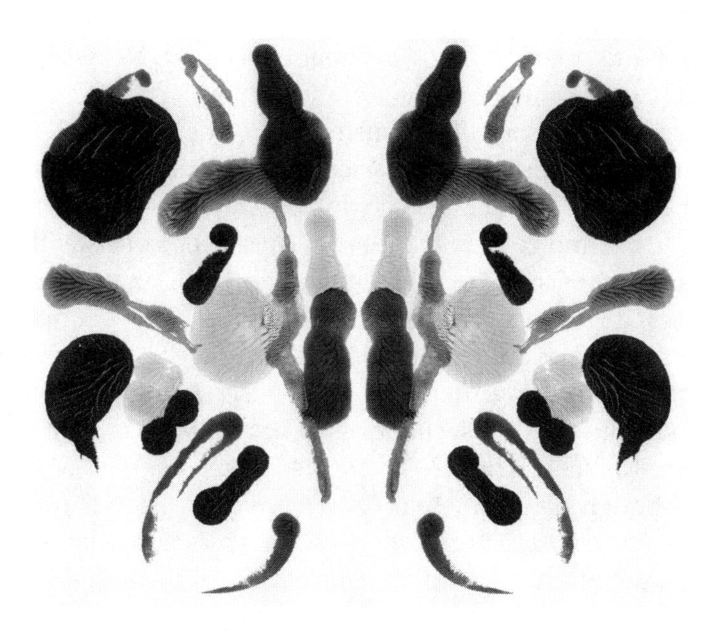

FIGURE 12.2 Inkblot.

movement and color (Rorschach, 1921). He does not appear to have been influenced by Freudian theories in constructing the inkblots or their interpretation, and instead had his own theory that the perception of movement and color would give insight into personality. In particular, he thought movement responses were related to introversion, whereas color responses were related to extraversion.

This idea that perception of movement and introversion were related appears partly based on muscle movement and dream research by a philosopher in the 1800s named John Mourly Vold (Ellenberger, 1993). Rorschach took Mourly Vold's idea that inhibition of movement during sleep would cause more dream imagery involving movement and applied it to the responses generated by his inkblots. In other words, his theory was that introverts should see more images that are moving in the blots, as a result of their being psychologically inhibited. Rorschach also outlined a theory that the perception and use of color in descriptions of the inkblots was related to affect and extraversion. In particular, those who used more color responses were more extraverted and likely to show high levels of emotion. Unlike his ideas about movement, however, his theory about color seems to have been pulled from common vernacular (e.g., "black moods") and personal opinion rather than any research or previous theories (Rapaport, Gill, & Shafer, 1946). Rorschach also seemed particularly interested in the balance of introversion and extraversion, called "Experience Balance" in English. The ratio of movement to color responses, he believed, would reveal a person's "basic experience and orientation toward reality" (Wood, Nezworski, Lilienfeld, & Garb, 2003, p. 25).

Rorschach's reasons for focusing on color and movement have not, though, been supported by the scientific evidence. Experience Balance, for example, has not consistently been demonstrated to be related to introversion or extraversion and color responses have not been consistently related to any particular diagnosis. It should be noted, however, that some of Rorschach's hypotheses do have some consistent support. For example, that a more intelligent person would provide higher numbers of movement responses has been supported to a moderate degree (see Frank [1979] for a review), as have some indicators of psychotic disorders (Lilienfeld, Wood, & Garb, 2001).

So, was Rorschach right? The answer is "mostly not" with the occasional "yes." Although his major hypotheses have not been shown to be correct, some minor ones have support. These inconsistencies and concerns led to numerous within-group conflicts during the 1930s and beyond, as different groups of researchers and clinicians developed further types of scores, or refined the meaning of certain scores. It was during these conflicts that some began to use the Rorschach as a more psychoanalytically oriented test, interpreting responses to blots as if they were dreams rather than relying on Rorschach's methods. At the same time, well-conducted research in the 1950s showed that the Rorschach was not more useful (and was in fact slightly less useful) than objective measures of personality and tended to overpathologize normal individuals (i.e., make them appear less healthy than they actually are). Further research showed that it added little to nothing in the way of diagnostic use if one already had access to biographical information and a person's history (see Garb [1998] for a review). By the beginning of the 1960s, most research-oriented and scientifically based psychologists thought the Rorschach was not a useful instrument.

However, a major reform attempt, one that likely saved the Rorschach from being consigned to the graveyard of psychological tests, was undertaken by John Exner, who developed his comprehensive system (CS; Exner, 2003). The CS included reviews of the literature, norms, and administration guidelines—all things that were lacking in manuals at the time. Exner also did both extensive research into reliability and validity of the traditional scores and developed new ones. At the same time, though, findings by researchers other than Exner or his associates began to appear, with results in sharp contrast to those reported in the CS's manual. In fact, the vast majority of the supportive studies cited in the latest CS manual are *unpublished* studies conducted by Exner and his research team at Rorschach Workshops.[3] Studies

[3] Recently, the majority of supportive studies for the Rorschach have been published in the *Journal of Personality Assessment*, a well-respected journal that publishes large amounts of high-quality research. It also happens to be the official journal of the Society for Personality Assessment, which originated as the Rorschach Institute, and is almost exclusively staffed by editors who are very strong proponents of the Rorschach's use.

conducted by those without ties to Exner showed identical problems as those raised in the 1950s and 1960s: overpathologizing, low diagnostic accuracy outside of psychotic disorders, lack of relationship to objective measures of psychopathology and personality (Hunsley & Bailey, 2001).

In summary, then, the Rorschach began life in 1922 as a theoretically shaky, nonempirically supported test for the majority of psychopathology (psychotic disorders being the exception). Despite almost 90 years of research, use, and various iterations of scoring and administration criteria, the evidence today indicates that it has changed little over the years. There are other, better assessment tools that should be used instead.

Thematic Apperception Test

The two major figures in the development of the thematic apperception test (TAT; Murray, 1943) were Henry Murray and Christina Morgan. Murray was a physician and biochemist before being hired to the faculty of the Harvard Psychological Clinic in 1926. Although largely unqualified for such a position initially, Murray underwent extensive training in psychoanalysis, including meeting with Carl Jung, and doing intensive reading in psychiatric and psychological literature (Robinson, 1992). Morgan was an artist and certified nurse's aide also highly influenced by Jung's theories on personality and psychopathology, having been analyzed by him personally (Douglas, 1993). Murray and Morgan may appear odd choices to develop a major psychological test, but the TAT ranks second only to the Rorschach as the most often used type of projective test by clinical psychologists (Camara, Nathan, & Puente, 2000).

Murray appears to have been the theoretical driving force behind the TAT, as it is based on his "needs-press" concepts of personality. For Murray, an individual's personality is the result of an interaction between one's needs (internal motivations) and presses (environmental or situational pressures that impact how one expresses those needs). Morgan assisted more in the preparation of the actual testing materials (the pictures on the test cards), some early administration of the measure, and writing the results for publication. The instrument itself has cards with black and white pictures of various kinds. Examiners show the cards to the examinee and ask him or her to tell a story based on the picture. The stories told, according to Murray, reveal numerous aspects of personality and can be used to understand how someone thinks and feels. Murray (1943, p. 1) believed that these stimuli would also "expose the underlying tendencies which the subject . . . is not willing to admit, or cannot admit because he is unconscious of them."

The TAT manuals provide very clear and detailed procedures, but similar to what happened with the Rorschach, numerous other systems and methods of using the TAT developed. The majority of practitioners do not appear to use any of the available scoring systems, though, instead relying on "intuitive" interpretations of the stories (Groth-Marnat, 2003). So, just as with the Rorschach, most users of the TAT are not using it as intended by the developers or even from the same theoretical viewpoint. There have been several positive findings regarding scoring on the TAT and relationship to specific areas of psychological functioning (e.g., personality disorders), but they have all been found when using a particular scoring system. With the majority of those using the TAT not using either standardized administration or scoring procedures, this information is a moot point. Add in the TAT's lack of incremental validity (Garb, 1998) and the high potential for overpathologizing normal populations based on TAT responses (Lilienfeld, Wood, & Grab, 2000), and it can be seen why the TAT "rarely plays a prominent role in clinical diagnostic evaluations" (Weiner & Greene, 2008, p. 469).

So, in summary, the TAT has some limited empirical support in assessing for personality disorders and achievement motives when using particular scoring systems. It does not, though, seem to be useful for broad psychopathology or personality assessment. This is especially true given that few practitioners use the TAT in the standardized manner that it was intended to be used, instead relying on personal experience and judgment, with all the attendant biases and problems relying on such entails (Dawes, Faust, & Meehl, 1989).

Figure Drawings

The final type of projective test to be discussed is not a specific measure, but instead a collection of measures. A number of methods to reportedly assess personality and psychopathology require that an individual draw pictures of a person, people, or objects. The three most widely used are the Draw-A-Person (DAP) test (Harris, 1963), the House–Tree–Person (HTP) test (Buck, 1948), and the Kinetic Family Drawing (KFD) test (Burns & Kaufman, 1970). They are in the top 15 most commonly used instruments by clinical and school psychologists, which is unsurprising given the speed and ease of their administration (many take fewer than 10 minutes).

Although each test has its own set of interpretation(s), there are two broad approaches to scoring figure drawings: the global approach and the sign approach (Lilienfeld et al., 2000; Weiner & Greene, 2008). In the global approach, interpretation is based on sets of indicators that are summed to yield a total score of adjustment (or lack thereof). The sign approach, in contrast, relies on identification of isolated features of the drawing (e.g., eye size, size of figure, placement of figure) that are supposedly related to specific pathology or personality problems. Constructing these drawings could purportedly bypass conscious efforts to hide or exaggerate symptoms and provide a more complete understanding of a person.

Large amounts of research over the last 60 years have been conducted to examine the reliability and validity of figure drawings, with highly varied results. Reliability of ratings in both sign and global approaches is variable and generally low. Validity studies across different projective drawings have met with a number of difficulties, particularly in the sign approach. A primary problem is lack of consistency in operational definitions. For instance, different studies or scoring systems often have the same feature interpreted in a different way. Some guidelines for interpreting drawings seem to almost specialize in making nonfalsifiable predictions. Hammer (1959) said that pathology could be seen in drawings that were too large or too small, lines that were too heavy or too light, and ones that had either too few or too many eraser marks. Others stated that those same signs could either indicate high levels of anxiety or successful coping efforts against high anxiety (Handler & Reyher, 1965). Or, it might be that, as Waehler (1997) contends, lack of validity in a drawing may simply occur because that individual does not show his or her distress in a drawing. Making such nonfalsifiable predictions and explaining away negative findings are hallmarks of pseudoscientific thinking, as we learned earlier.

Specific research examining the validity of the sign approach for different psychological characteristics shows the problems one would expect based on the previous information. Reviews of the KFD concluded that individual signs showed little to no relation to actual psychopathology (Handler & Habenicht, 1994). A study examining depressive and anxious symptoms in children on an inpatient psychiatric ward used both projective measures and objective measures (Joiner, Schmidt, & Barnett, 1996). It is interesting to note that this study found that the differing projective measures not only did not relate to scores on the objective measures, but also did not have a relationship to scores from the different projective measures (even another drawing measure!).

Despite the lack of validity demonstrated by the sign approaches, however, there is a silver lining for projective drawings. In a study examining the KFD and DAP, Tharinger and Stark (1990) were able to accurately distinguish between children with and without mood disorders, whereas the DAP distinguished among children who had mood disorders and mixed mood/anxiety problems. Further, there has been some support for the use of another global scoring procedure for the DAP, the screening procedure for emotional disturbance, to differentiate between groups of children with and without disruptive behavior problems (Naglieri & Pfeffier, 1992). However, other researchers found much lower effect size differences and concluded that it was of limited utility in the schools (Wrightson & Saklofske, 2000).

Even these positive findings, though, must be interpreted cautiously at this point. One reason is that it is not known whether controlling for intelligence, which has been shown to be lower across many types of psychopathology, would reduce or eliminate the positive findings reviewed previously. In other words, research needs to be done to rule out these rival hypotheses. The lone study that addressed that issue (Schneider, 1978) found that controlling for intelligence eliminated the possible incremental validity of drawings given to school-age children when assessing for behavior problems. The complex role of artistic ability in impacting scores and interpretations is also not well understood (Lilienfeld et al., 2000). Also problematic is the fact that it is unknown how many practicing clinicians use a sign versus a global approach, although a small study of active practitioners (Smith & Dumont, 1995) suggests that the vast majority of those who rely on drawings for clinical hypotheses use some combination of the approaches.[4]

In summary, it does appear that there may be limited uses for global scoring systems for projective drawings, in particular using the DAP and KFD for assessment of general behavioral and mood problems. There are not, however, any well-replicated lines of research that support the use of projective drawings and interpretation to differentiate children or adults for specific disorders. Further research on this issue, particularly as regards global scoring systems, should be conducted.

Conclusions About Projectives

After looking at the actual evidence, several things have become apparent. First, not all projective tests or techniques were created, or have been researched, equally (not dissimilar to what we saw in the previous chapter regarding herbal medicines). In their theoretical constructs, intended uses, and research-supported uses, they differ greatly. Second, the use of projective methods is not an either-or proposition. In opposition to the beliefs of their staunchest supporters, they are not empirically supported to be equally adept at assessing all aspects of personality and psychopathology. And in opposition to the beliefs of their staunchest critics, the research evidence does support the use of projective measures for assessing specific psychological constructs. Although certain measures have been useful in measuring psychotic disorders (Rorschach), personality disorders (TAT), disruptive behavior and mood problems (global figure drawing scores), those are the extent of their evidence base and they shouldn't be used outside of those small areas. In contrast, objective measures (such as the Behavior Assessment System for Children, the Minnesota Multiphasic Personality Inventory, or the NEO Personality Inventory) measure such constructs and many, many more accurately, making them a much better bang for your assessment buck.

---

[4] The first author's (CWL) father uses a metaphor to describe what happens when mixing things that shouldn't go together. He says, "You know what you get when you mix a gallon of ice cream with a gallon of manure? Two gallons of manure!" (C. W. Lack, personal communication, July 2012).

HOW TO CHOOSE A GOOD MENTAL HEALTH PROVIDER

Given the preponderance of non-EBT for psychological problems, one must often be careful in choosing a provider of mental health services, whether that person is a psychologist, psychiatrist, professional counselor, or other kind of therapist. The best advice we can give anyone when choosing a mental health professional is to see someone who practices evidence-based psychology. As discussed at the start of this chapter, EBP is a guiding principle that means a clinician is guided in the treatment and assessment methods he or she uses by current best practices, as defined by scientific evidence. It is unfortunate that many therapists have not been trained in these methods, and instead rely on intuition, what they think has worked well, or what they were trained in—regardless of the evidence or lack thereof for it's effectiveness. Asking potential therapists what their primary therapeutic orientation is and how they know the type of therapy they do works, are great ways to find out whether therapists use EBP.

Our second piece of advice is that people need to be sure that the mental health provider does not attempt to push his or her own personal values system onto you, the client. Although this is both an unethical and inappropriate thing to do, from the first author's (CWL) own experience with clients and working with several national organizations, a large number of clients report this as happening to them. This does not mean that you need to find a therapist with your exact religious, political, ethnic, and cultural background, but it does mean that your therapist needs to respect what your beliefs and values are and recognize that his or her job as a therapist is not to convert you to a different belief system. If you find yourself in a situation in which this is occurring, we would recommend giving the therapist a warning that you are becoming offended by such actions. If he or she continues to push an agenda at the expense of your mental health, a report to the state licensing board would be appropriate.

CONCLUSIONS

Psychological science has advanced immensely in the 130 years or so since its inception, both in terms of basic knowledge and applied aspects such as the assessment and treatment of mental health problems. Just as is the case when seeking out medical treatments, those who need mental health services should be able to have access to an evidence-based practitioner in order to maximize their chances for symptom reduction or an accurate diagnosis. Unfortunately, given the gap that exists between what we know actually works and what many practitioners use on a daily basis, a large amount of the onus for accessing such care still falls on the consumer. In this, as in so many other areas of life, your ability to think critically and evaluate the evidence will be a major boon.

QUESTIONS FOR REFLECTION

1. *Given how thoroughly discredited Andrew Wakefield's research on vaccines and autism is, how might we explain the ongoing fascination with the notion that vaccines can cause autism? Do you know anyone who thinks there is a link? What reasons has this person given you?*
2. *Why is AC more respectable and evidence based than FC?*
3. *What does the idea of "purple hat therapy" tell us about how to interpret claims made in mental health treatment? What about for various forms of complementary and alternative medicine?*
4. *Does the popularity of AA do more harm than good in terms of getting in the way of secular, evidence-based therapy and treatment?*
5. *The Rorschach test is not only empirically shaky, but also lends itself to abuse. What sorts of harm can result from overconfidence in the usefulness of this and similar tests?*

REFERENCES

Alcoholics Anonymous. (2014). *A.A. membership survey*. Retrieved from http://www.aa.org/assets/en_US/aa-literature/p-48-aa-membership-survey

Amaya-Jackson, L., Reynolds, V., Murray, M., McCarthy, G., Nelson, A., Cherney, M., . . . March, J. (2003). Cognitive behavioral treatment for pediatric posttraumatic stress disorder: Protocol and application in school and community settings. *Cognitive and Behavioral Practice, 10*, 204–213.

American Psychiatric Association. (2013). *Diagnostic and statistical manual of mental disorders* (5th ed.). Arlington, VA: American Psychiatric Press.

American Speech-Language-Hearing Association. (1994). *Facilitated communication* [Technical Report]. Available from http://www.asha.org/policy/TR1994-00139.htm

APA Presidential Task Force on Evidence-Based Practice. (2006). Evidence-based practice in psychology. *American Psychologist, 61*, 271–285.

Atwood, K. C., Woeckner, E., Baratz, R. S., & Sampson, W. I. (2008). Why the NIH Trial to Assess Chelation Therapy (TACT) should be abandoned. *Medscape Journal of Medicine, 10*(5), 115.

Bland, S., Valoroso, L., Stranges, S., Stazzullo, P., Farinaro, E., & Trevisan, M. (2005). Long-term follow-up of psychological distress following earthquake experiences among working Italian males: A cross-sectional analysis. *Journal of Nervous & Mental Disease, 193*, 420–423.

Buck. J. N. (1948). The H-T-P technique, a qualitative and quantitative method. *Journal of Clinical Psychology, 4*, 317–396.

Burns, R. C., & Kaufman, S. H. (1970). *Kinetic family drawings* (K-F-D)*: An introduction to understanding children through kinetic drawings*. New York, NY: Bruner/Mazel.

Butcher, J. N., Butcher, J., Tellegen, A., Graham, J., & Graham J. R. (1989). Minnesota Multiphasic Personality Inventory®-2 (MMPI®-2). San Antonio, TX: Pearson.

Callahan, R., & Callahan, J. (1996). *Thought field therapy (TFT) and trauma: Treatment and theory*. Indian Wells, CA: Thought Field Therapy Training Center.

Camara, W. J., Nathan, J. S., & Puente, A. E. (2000). Psychological test usage: Implications in professional psychology. *Professional Psychology: Research and Practice, 31*, 141–154.

Chambless, D. L., Sanderson, W. C., Shoham, V., Bennett Johnson, S., Pope K. S., Crits-Christoph, P., . . . McCurry, S. (1996). An update on empirically validated therapies. *Clinical Psychologist, 49*(2), 5–18.

Connors, G. J., & Dermen, K. H. (1996). Characteristics of participants in secular organizations for sobriety. *American Journal of Drug and Alcohol Abuse, 22*(2), 281–295.

Costa, P. T., Jr. & McCrae, R. R. (2010). NEO Personality Inventory–3 (NEO-PI-3). Port Huron, MI: Sigma Assessment Systems, Inc.

Craig, G. (Producer). (1997). *Six days at the VA: Using emotional freedom therapy* [Videotape]. Available from Gary Craig, 1102 Redwood Blvd, Novato, CA 94947.

Dawes, R. M., Faust, D., & Meehl, P. E. (1989). Clinical versus actuarial judgment. *Science, 243*, 1668–1674.

Douglas, C. (1993). *Translate this darkness: The life of Christiana Morgan*. New York, NY: Simon & Schuster.

Ehlers, A., Mayou, R. A., & Bryant, B. (1998). Psychological predictors of chronic posttraumatic stress disorder after motor vehicle accidents. *Journal of Abnormal Psychology, 107*, 508–519.

Ellenberger, H. F. (1993). *Beyond the unconscious*. Princeton, NJ: Princeton University Press.

Exner, J. E. (2003). *The Rorschach: A comprehensive system: Basic foundations* (Vol. 1, 4th ed.). Hoboken, NJ: John Wiley & Sons.

Ferri, M., Amato, L., & Davoli, M. (2006). Alcoholics anonymous and other 12-step programmes for alcohol dependence. *Cochrane Database of Systematic Reviews, 3*, CD005032.

Frank, G. (1979). On the validity of hypotheses derived from the Rorschach: VI. *M* and the intrapsychic life of individuals. *Perceptual and Motor Skills, 48*, 1267–1277.

Frank, L. (1939). Projective methods for the study of personality. *Journal of Psychology, 8*, 389–413.

Garb, H. N. (1998). *Studying the clinician: Judgment research and psychological assessment*. Washington, DC: American Psychological Association.

Groth-Marnat, G. (2003). *Handbook of psychological assessment* (4th ed.). Hoboken, NJ: John Wiley & Sons.

Guadino, B. A., & Herbert, J. D. (2000). Can we really tap our problems away? A critical analysis of thought field therapy. *Skeptical Inquirer, 24*, 29–33.

Hammer, E. F. (1959). Critique of Swenson's "empirical evaluation of human figure drawings." *Journal of Projective Techniques, 23*, 30–32.

Handler, L., & Habenicht, D. (1994). The kinetic family drawing: A review of the literature. *Journal of Personality Assessment, 62*, 440–464.

Handler, L., & Reyher, J. (1965). Figure drawing anxiety indices: A review of the literature. *Journal of Projective Techniques, 29*, 305–313.

Harris, D. B. (1963). *Children's drawings as measures of intellectual maturity: A revision and extension of the goodenough draw-a-man test*. Orlando, FL: Harcourt Brace.

Hojnoski, R. L., Morrison, R., Brown, M., & Matthews, W. J. (2006). Projective test use among school psychologists: A survey and critique. *Journal of Psychoeducational Assessment, 24*, 145.

Hunsley, J., & Bailey, J. M. (1999). The clinical utility of the Rorschach: Unfulfilled promises and uncertain future. *Psychological Assessment, 11*, 266–277.

Hunsley, J., & Bailey, J. M. (2001). Whither the Rorschach? *Psychological Assessment, 13*, 472–485.

Jain, A., Marshall, J., Buikema, A., Bancroft, T., Kelly, J. P., & Newschaffer, C. J. (2015). Autism occurrence by MMR vaccine status among US children with older siblings with and without autism. *Journal of the American Medical Association, 313*(15), 1534–1540.

Jaycox, L. H., Tanielian, T. L., Sharma, P., Morse, L., Clum, G., & Stein, B. D. (2007). Schools' mental health responses after Hurriances Katrina and Rita. *Psychiatric Services, 58*, 1339–1343.

Joiner, T. E., Schmidt, K. L., & Barnett, K. (1996). Size, detail, and line heaviness in children's drawings as correlates of emotional distress: (More) negative evidence. *Journal of Personality Assessment, 67*, 127–141.

Kane, K. (2006, January 18). Drug error, not chelation therapy, killed boy, expert says. *Pittsburgh Post-Gazette*. Retrieved from www.postgazette.com/pg/06018/639721.stm

Kelly, J. F., Magill, M., & Stout, R. L. (2009). How do people recover from alcohol dependence? A systematic review of the research on mechanisms of behavior change in alcoholics anonymous. *Addiction Research & Theory, 17*(3), 236–259.

Lack, C. W. (2008). *Tornadoes, children, and posttraumatic stress*. Saarbrücken, Germany: VDM Verlag Dr. Müller.

Lambert, M. J., & Ogles, B. M. (2004). The efficacy and effectiveness of psychotherapy. In M. J. Lambert (Ed.), *Bergin and Garfield's handbook of psychotherapy and behavior change* (5th ed., pp. 139–193). Hoboken, NJ: Wiley.

Lilienfeld, S. O., Wood, J. M., & Garb, H. N. (2000). The scientific status of projective techniques. *Psychological Science in the Public Interest, 1*, 27–66.

Lilienfeld, S. O., Wood, J. M., & Garb, H. N. (2001). What's wrong with this picture? *Scientific American, 284*(5), 80–87.

Lohr, J. M., Gist, R., Deacon, B., Devilly, G. J., & Varker, T. (2015). Science and non-science-based treatments for trauma-related stress disorders. In S. O. Lilienfeld, S. J. Lynn, & J. M. Lohr (Eds)., *Science and pseudoscience in clinical psychology* (2nd ed, pp. 277–321). New York, NY: Guilford Press.

MacKillop, J., & Gray, J. C. (2014). Controversial treatments for alcohol use disorders. In S. O. Lilienfeld, S. J. Lynn, & J. M. Lohr (Eds.), *Science and Pseudoscience in Clinical Psychology* (2nd ed., pp. 322–363). New York, NY: Guilford Press.

McNally, R. J. (1999). On eye movements and animal magnetism: A reply to Greenwald's defense of EMDR. *Journal of Anxiety Disorders, 13*, 617–620

McNally, R. J., Bryant, R. A., & Ehlers, A. (2003). Does early psychological intervention promote recovery from posttraumatic stress? *Psychological Science in the Public Interest, 4*(2), 45–79.

Mitchell, J. T., & Everly, G. S., Jr. (1998). Critical incident stress management: A new era in crisis intervention. *Traumatic Stress Points, 12*, 6–11.

Morganthaler, W. (1954). Der kampf um das erscheinen der Psychodiagnostik [The struggle for the publication of *Psychodiagnostics*]. *Zeitschrift eur diagnostiche psychologie und perseonlichkeitsforschung, 2*, 255–271. [An English translation is available upon request from James M. Wood, jawood@utep.edu.]

Murray, H. A. (1943). *Thematic apperception test manual*. Cambridge, MA: Harvard University Press.

Naglieri, J. A., & Pfeffier, S, I. (1992). Performance of disruptive behavior-disordered and normal samples on the Draw-A-Person: Screening procedure for emotional disturbance. *Psychological Assessment, 4*, 156–159.

National Institute on Drug Abuse. (2015). *Trends & statistics*. Retrieved from http://www.drugabuse.gov/related-topics/trends-statistics

Polenick, C. A., & Flora, S. R. (2012). Sensory integration and autism: Science or pseudoscience? *Skeptic, 17*(2), 28–35.

Rapaport, D., Gill, M., & Schafer, R. (1946). *Diagnostic psychological testing* (Vol. 1). Chicago, IL: Year Book.

Reynolds, C. R., & Kamphaus, R. W. (2015). Behavior Assessment System for Children, Third Edition (BASC-3). San Antonio, TX: Pearson.

Robinson, G. E. (1992). *Love's story told: A life of Henry A. Murray*. Cambridge, MA: Harvard University Press.

Romancyzk, R. G., Turner, L. B., Sevlever, M., & Gillis, J. M. (2015). The status of treatment for autism spectrum disorders: The weak relationship of science to interventions. In S. O. Lilienfeld, S. J. Lynn, & J. M. Lohr (Eds.), *Science and pseudoscience in clinical psychology* (2nd ed., pp. 431–465). New York, NY: Guilford Press.

Rorschach, H. (1921). *Psychodiagnostics: A diagnostic test based on perception*. New York, NY: Grune & Stratton.

Rosen, G. M., & Davison, G. C. (2003). Psychology should list empirically supported principles of change (ESPs) and not credential trademarked therapies or other treatment packages. *Behavior Modification, 27*, 300–312.

Sackett, D. L., Rosenberg, W. M., Gray, J. A. M., Haynes, R. B., & Richardson, W. S. (1996). Evidence based medicine: What it is and what it isn't. *British Medical Journal, 312*, 71–72.

Schneider, G. B. (1978). A preliminary validation study of the Kinetic School Drawing. *Dissertation Abstracts International, 38*(11-A), 6628.

Shapiro, F. (1995). *Eye movement desensitization and reprocessing: Basic principles, protocols, and procedures*. New York, NY: Guilford Press.

Smith, D., & Dumont, F. (1995). A cautionary study: Unwarranted interpretations of the Draw-A-Person test. *Professional Psychology: Research and Practice, 26*, 298–303.

Spates, C. R., Koch, E., Cusack, K., Pagoto, S., & Waller, S. (2009). Eye movement desensitization and reprocessing. In E. B. Foa, T. M. Keane, M. J. Friedman, & J. A. Cohen (Eds.), *Effective treatments for PTSD: Practice guidelines from the International Society for Traumatic Stress Studies*. New York, NY: Guilford Press.

Stewart, P. A., Hyman, S. L., Schmidt, B. L., Macklin, E. A., Reynolds, A., Johnson, C. R., . . . Manning-Courtney, P. (2015). Dietary supplementation in children with autism spectrum disorders: Common, insufficient, and excessive. *Journal of the Academy of Nutrition and Dietetics, 115*, 1237–1248.

Tharinger, D. J., & Stark, K. (1990). A qualitative versus quantitative approach to evaluating the Draw-A-Person and kinetic family drawing: A study of mood and anxiety-disordered children. *Psychological Assessment, 2*, 365–375.

U.S. Department of Labor. (2010). *Occupational Outlook Handbook; 2010–2011 Library Edition, Bulletin 2800*. Washington, DC: Superintendent of Documents, U.S. Government Printing Office.

Waehler, C. A. (1997). Drawing bridges between science and practice. *Journal of Personality Assessment, 69*, 482–487.

Waite W. L., & Holder, M. D. (2003). Assessment of the emotional freedom technique: An alternative treatment for fear. *Scientific Review of Mental Health Practice, 2*(2), 20–26.

Walters, G. D. (2000). Spontaneous remission from alcohol, tobacco, and other drug abuse: Seeking quantitative answers to qualitative questions. *American Journal of Drug and Alcohol Abuse, 26*(3), 443–460.

Weiner, I. B., & Greene, R. L. (2008). *Handbook of personality assessment*. Hoboken, NJ: John Wiley & Sons.

Weitlauf, A. S., McPheeters, M. L., Peters, B., Sathe, N., Travis, R., Aiello, R., . . . Warren, Z. (2014). *Therapies for children with autism spectrum disorder: Behavioral interventions update*. Rockville, MD: Agency for Healthcare Research and Quality. Retrieved from http://www.ncbi.nlm.nih.gov/books/NBK241444

Wood, J. M., Nezworski, M. T., Lilienfeld, S. O., & Garb, H. N. (2003). *What's wrong with Rorscach? Science confronts the controversial inkblot test*. New York, NY: Jossey-Bass.

Work Group on ASD and PTSD. (2004). *Practice guideline for the treatment of patients with acute stress disorder and posttraumatic stress disorder*. Arlington, VA: American Psychiatric Press.

World Health Organization. (2014). *Global status report on alcohol and health 2014*. Retrieved from http://apps.who.int/iris/bitstream/10665/112736/1/9789240692763_eng.pdf

Wrightson, L., & Saklofske, D. H. (2000). Validity and reliability of the Draw A Person: Screening procedure for emotional disturbance with adolescent students. *Canadian Journal of School Psychology, 16*(1), 95–102.

CHAPTER 13

THE RELATIONSHIP BETWEEN SCIENCE AND RELIGION

Despite having its origins in the work of classical Greek philosophers, such as Thales and Aristotle, modern science and scientific thinking are relatively recent arrivals, dating back only a few hundred years. As such, when it comes to influencing human thought and culture, scientific thinking is very much the new kid on the block as compared to religion. Religious practices, after all, have been convincingly traced back to the Upper Paleolithic era of 30,000 years ago, with some scholars dating religious or proto-religious behaviors back another 70,000 years or so (Bellah, 2011). Today, there is an increasingly loud chorus of voices from many spectrums of thought that are pushing the idea of a "culture war" between science or scientific skepticism and religion. The interplay between these two modes of thought has a huge influence on the culture at large, as does people's beliefs about the relationship between the two. To move forward in understanding this complex relationship, though, we need to operationalize what we mean by science, religion, and culture.

At this point in the book you are undoubtedly familiar with what science is and is not. But, just in case you need a refresher, we have been defining *science* as "A set of methods designed to describe and interpret observed or inferred phenomena, past or present, and aimed at building a testable body of knowledge open to rejection or confirmation" (Shermer, 2002, p. 18). As discussed extensively in Chapter 1, science strives to be empirical, changes in response to new information, and is naturalistic in its view of the world.

Common definitions of religion tend to emphasize three qualities: substantive (what it concerns—the sacred or supernatural), functional (what you do as a part of your religion), and personal (an individual's belief, emotion, and behavior tied to that religion). A definition by Koenig (2008) illustrates the substantive and functional aspects:

> A system of beliefs and practices observed by a community, supported by rituals that acknowledge, worship, communicate with, or approach the Sacred, the Divine, God (in Western cultures), or Ultimate Truth, Reality, or nirvana (in Eastern cultures). (p. 11)

Complementary to this is the definition used by James (1902), which illustrates the individual aspects of religion: "the feelings, acts, and experiences of individual men in their solitude, so far as they apprehend themselves to stand in relation to whatever they may consider the divine" (p. 42). Given the wide variety of beliefs and behaviors that people who are ostensibly of the same religion can evidence, keeping this individualistic aspect in mind is key when discussing multiple issues later in the chapter.

As given previously, the definitions of science and religion have differences and similarities. Broadly speaking, science concerns itself with the natural, religion with the supernatural. Science focuses on testable claims and hypotheses, whereas religion focuses on individual beliefs. Both, though, place an emphasis on a way to know or understand the world, even as these ways vary dramatically. For science, it is through a systematic procedure that is designed to test your ideas thoroughly and critically, to make sure that they are accurate. For religion, understanding the world is most often done via divine revelation, be it through a sacred text or individual prayer. The rate of change seen in these two fields is also vastly different. Science, optimally, both changes rapidly and welcomes change; this means our knowledge about a given topic is advancing with the rejection of false hypotheses. Religious beliefs tend to change much more slowly, especially if they are based on some sort of sacred text or revelation. Given these differences, it is quite understandable that there could be some conflicts that emerge between not just these worldviews, but also the people who hold them. These conflicts are most likely to occur in areas of overlap between the two, where both are attempting to explain some particular phenomenon or aspect of the world.

Much like with science and religion, there is no single, agreed-upon definition of culture. Generally, culture can be thought of as the behaviors, values, beliefs, and worldview shared by a group of people and which are transmitted from generation to generation. This may include attitudes about and behaviors toward out-group members, laws and customs, moral values, artistic and literary styles, knowledge, and all other aspects that make you a member of "us" and people from a different culture a member of "them." On a deep level, the culture a person is a part of influences basic assumptions about human relationships, nature, and activity, as well as how one views reality. For many people, culture is highly influenced by the dominant (and nondominant) religion in their region. In turn, religion is in many ways shaped by the dominant culture around it. Both, in modern society, are also influenced by science, both in terms of basic discoveries that build theory and more applied aspects (such as technology or medicine). The interplay among these three (science, religion, and culture) is complex, highly influential on most people's lives, and the subject of this, the penultimate chapter of our book.

RELATIONSHIPS BETWEEN SCIENCE AND RELIGION

Many people have written on the potential ways that science and religion can interact (see Haught [1995] or Polkinghorne [1998] for more in-depth examinations than this chapter allows). How a culture views the relationship between them can have massive implications on social, political, educational, and personal levels. These interactions are typically divided into four types of perspectives:

- Incompatibility and conflict
- Independence and contrast
- Dialogue and contact
- Integration and confirmation

The view that science and religion are inherently at odds with one another (*incompatibility and conflict*) promotes the notion that there will always be conflict between the ideas of religion and the ideas of science because the two are mutually incompatible. Proponents of this view typically point to the sharp contrast between the supernatural and naturalistic views of the world and the "ways of knowing" that the two have. Scientific empiricism, with its reliance on experimental replication and verification, is seen as directly opposed to faith, which relies on believing despite a lack of evidence (i.e., of the scientific kind). Today, many prominent fundamentalist religious leaders (James Dobson, David Barton), philosophers (Daniel Dennett), and scientists (Neil deGrasse Tyson, Richard Dawkins, Lawrence Krauss) hold to the incompatibility hypothesis. A brief scan of the past 600 years of Western history reveals several areas of overlap that caused conflicts between religious institutions and scientific knowledge, from the age of the universe to human sexuality to evolution via natural selection.

A different viewpoint on the relationship between science and religion stresses *independence and contrast*. Popularized by Gould (1999), an evolutionary biologist and science writer, this view sees religion and science as dealing with fundamentally different domains or aspects of the universe and human existence. These "nonoverlapping magisteria" (NOMA), as Gould called them, can allow for science and religion to exist without conflict, as long as each stays on its side of the fence:

> Science tries to document the factual character of the natural world, and to develop theories that coordinate and explain these facts. Religion, on the other hand, operates in the equally important, but utterly different, realm of human purposes, meanings, and values—subjects that the factual domain of science might illuminate, but can never resolve. (p. 4)

Another way to think about this is that religion should confine itself to investigating and discussing what we should or ought to do (a prescriptive view), whereas science sticks to how the universe actually is (a descriptive view). Supporters of this view argue that when science becomes prescriptive, or religion attempts to be descriptive, conflicts can arise (as we will see later).

The NOMA view has received both strong support and strong criticism. For example, the National Academy of Science's official position is that science cannot investigate the supernatural (i.e., religious claims) because it concerns itself with only the natural, so no conflict between religion and science has to exist (National Academy of Sciences & Institute of Medicine [NAS & IOM], 2008). Charles Darwin himself may have fallen into this camp, as he wrote in a letter from 1879 that "Science has nothing to do with Christ, except insofar as the habit of scientific research makes a man cautious in admitting evidence" (Darwin, 1887, p. 307). In contrast, numerous individual theologians, scientists, and philosophers reject this attempt at compromise. For example, Richard Dawkins, Jerry Coyne, and others see numerous religious beliefs as being testable via science, such as the view held by biblical literalists that the entire universe was created in the relatively recent past (6,000–10,000 years ago), which contrasts starkly with robust scientific evidence showing the universe to be approximately 13.8 billion years old.

The "science and religion" movement emphasizes *dialogue and contact*, saying that science and religion should work with each other, rather than be at odds or studying different areas. Most in this group tend to be either religious scientists or scientifically sympathetic religionists. For example, Francis Collins is an American physician, geneticist, and director of the National Institutes of Health who also describes himself as a deeply committed, evangelical Christian. He started the BioLogos foundation, a group dedicated to bridging gaps between science and religion. The BioLogos mission "invites the church and the world to see the harmony between science and biblical faith as we present an evolutionary understanding of God's creation" (BioLogos Foundation, n.d.). Such attempts have also been extolled by the Catholic Church, whose official doctrine embraces divinely guided evolutionary processes. The National Center for Science Education views this movement as being potentially beneficial to the public understanding of science, but cautions not to confuse this with "theistic science" (Scott, 1998).[1]

The final view, which is that of *integration and confirmation*, focuses on the very complex but generally peaceful relationship between science and religion. In it, the historical relationship between different religions and the search for truth via empirical investigation is emphasized. Rather than conflicting or separate domains of knowledge, the vast majority of

[1] Theistic science proponents (such as Alvin Plantinga or Steven Meyer) aim to change the way science has been practiced by allowing for supernatural injections into the scientific process (e.g., "I'm not sure why this is, so God must have done it"), rather than relying on purely naturalistic ones.

scientific exploration on this view has been undertaken and maintained by the religious communities of the time. Egyptian, Greek, Christian, Muslim, Buddhist, and Hindu institutions have all integrated and supported the vast majority of scientific discoveries, particularly those focused on technological advances, via universities and other centers of learning. This interdependence has been the norm, with conflicts arising on a particular few topics related to "ultimate questions" such as how life began (Fara, 2009). Today, for example, those who reject the idea that humans evolved across millions of years from other, similar hominids (as supported by the theory of evolution via natural selection) have no problem accepting the idea that objects are pulled toward those with large amounts of mass because of gravity (as supported by the theory of general relativity).

Each of these views has its supporters and detractors. Probably the loudest voices are those promoting the incompatibility and conflict model, which can be traced back to the 1800s. For example, Andrew D. White (cofounder and first president of Cornell University) published a two-volume tome on the history of conflict between modern science and Christianity (showcasing disagreements over topics in biology, astronomy, geography, psychology, archeology, and more). In it, he showed how some of these battles can take decades to centuries to reconcile, but argued that there was a relatively predictable pattern to the resolution (White, 1896). Although he used the Christian church, parts of the world with other dominant religions (e.g., Islam) often show a similar pattern.

1. An individual or group proposes a new idea that conflicts with traditional religious doctrine or beliefs, but this tends to be ignored by the church.
2. The idea becomes increasingly accepted by others, spreading and gaining momentum.
3. The church condemns the idea, typically using passages from the Old and New Testament to justify why the new idea is wrong.
4. The idea gains more and more support from the public and noted intellectuals.
5. The church puts out a statement showing how the idea either attacks a fundamental principle of Christian belief or even tries to refute the underpinnings of the entire religion.
6. Support continues to grow among the public.
7. Churches and church members begin to ignore the idea, and subsequently may ignore the biblical verses it used in step 3 to condemn the idea.
8. The church eventually incorporates the idea into its belief system, although this may take decades to centuries.[2]

[2] Much of White's scholarship is today considered shoddy (e.g., saying that most learned people believed the Earth was flat in the Middle Ages, when this was not true) and highly influenced by his own beliefs (which were decidedly antireligious). Still, the cycle that he identified serves as a useful model for showing how science, religion, and culture interact to change one another.

A model conflict that followed these steps involved a little planet we call Earth, and whether or not it was stationary or orbited around the sun. In 1543, Nicolaus Copernicus published a book called *On the Revolutions of the Celestial Spheres*, in which he outlined a model of the universe that placed not the Earth, but the sun at the center of all things. His heliocentric (or sun-centered) theory of the universe was in direct conflict with the teachings of the Catholic Church at that time, which used a literal interpretation of certain passages in the Bible to support the notion of an immobile Earth. Although Copernicus did not live to see it, having died as the book was being published, church officials soon denounced his ideas. This served to stifle more research in this area across much of mainland Europe, although an English astronomer named Thomas Digges was able to advance and progress Copernicus's ideas, given the relative lack of Catholic influence in Britain.

The deliberate suppression of the ideas of a heliocentric universe, with its challenge to deeply held religious beliefs, came to a head in the first decades of the 1600s with an individual named Galileo Galilei. Using a newly developed scientific tool, the telescope, Galileo was able make observations that supported a heliocentric, rather than geocentric, solar system. He published his findings in 1610 and, unlike Copernicus 60 years earlier, was met with a highly hostile response from the Catholic Church. A formal decree was issued in 1616 that banned any books that espoused the idea of heliocentrism and Earth's mobility, labeling the idea as formally heretical. It declared that such an idea was "foolish and absurd in philosophy, and formally heretical since it explicitly contradicts in many places the sense of Holy Scripture." Further, Galileo was specifically ordered to not teach, discuss, or write about heliocentrism.

Not one to just give up, Galileo continued to write and argue for the idea of a moving Earth, culminating in his publishing a hugely popular book called *Dialogue Concerning the Two Chief World Systems* in 1632. In it, he sneakily promoted heliocentrism through a series of dialogues among three characters. This angered many in the church, with Galileo being called to stand trial for heresy and disobeying the order to not promote or teach heliocentric views. As a result of this trial, not only was his major work banned, but the 69-year-old scientist was also sentenced to life imprisonment. This was mercifully commuted to house arrest by the next year, under which Galileo was kept until his death in 1642. It was not until 1835, almost two centuries after his death, that the Catholic Church removed Copernicus's *Revolutions* and Galileo's *Dialogues* from its *Index Librorum Prohibitorum*, a list of banned books that promoted heretical ideas. Over 150 years after that, in 1992, Pope John Paul II formally acknowledged the wrongs committed by the Church against Galileo and his ideas.

Although many see the Galileo story as emblematic of the inherent and intractable conflict between science and religion, the truth is much more complex. For instance, one intriguing aspect about the heliocentrism affair is that we are not talking about nontheistic, unbelieving scientists speaking out against church doctrine. Instead, Copernicus and Galileo were both devout Christians who respected the religious institutions of their time and place. Like almost all natural philosophers and scientists prior to the 1800s, Galileo was motivated to integrate his scientific findings with his theological beliefs, not trying to disprove any particular religion or religious belief. In fact, some see this whole matter not as Christianity conflicting with or suppressing science, but instead as a conflict between conservative (Catholic Church of the early 1600s) and liberal (Galileo) viewpoints on how biblical scripture should be interpreted (Lindberg & Numbers, 1987). This story—of religious scientists making discoveries that they tried to integrate with their faith, sometimes successfully, sometimes not—is actually the story of most of science prior to the 20th century.

In fact, dating back to early writings in the Christian faith, theologians often highly valued those who studied the natural world, believing that they were learning more about the world their God created. As outlined in the 16th-century Belgic Confession, the overwhelming majority of religious leaders, lay people, and most scientists prior to the 19th century saw the natural world and their sacred texts as being like two books written by the same author. For example, several early geologists who pushed back the age of the Earth into the millions of years, rather than thousands of years as typically held by religious thought at that point, were also clergymen, such as Adam Sedgwick. These findings did not diminish their religious faith—they instead shifted their faith to accommodate these findings. Charles Hodge as quoted in Noll (1994, p. 183), one of the most important theologians of the 1800s and a huge influence on modern evangelical thought, wrote, "Nature is as truly a revelation of God as the Bible; and we only interpret the Word of God by the Word of God when we interpret the Bible by science."[3] Although there have been and continue to be conflicts between religion and science, it turns out that this has not been a "default" position throughout history on most issues.

SCIENCE, RELIGION, AND BELIEFS IN THE MODERN WORLD

That science and religion have worked in concord, both in society at large and in the minds of individual scientists, has been shown repeatedly though history, but what about in modern times? As reviewed previously, there are

[3] It must be noted, though, that Hodge wrote that in 1859, before Darwin published his theory of evolution via natural selection. Later in life, Hodge would viciously attack Darwin's ideas as promoting atheistic views and actively lobby against their teaching in universities.

prominent scientists, philosophers, and religious leaders who fit into all of the categories detailed in the previous section, although the conflict model, pitting religion against science, has probably the loudest and most vocal group. One of the issues currently generating the most acrimony is that of origins science, which attempts to answer the big question of "How did this all start?" and typically involves the theories of evolution via natural selection and the Big Bang cosmological model. There are many religiously influenced groups that actively oppose the teaching of evolution in schools, or want to supplement it by teaching theistic creation as a scientifically valid alternative called "intelligent design."

One should note that most religious institutions in the United States do not have official doctrines that are opposed to the scientific views on evolution or the Big Bang. A large-scale survey recently found that only 11% of the U.S. population belongs to a religious denomination that has a belief system that explicitly rejects the theory of evolution via natural selection and/or the Big Bang (Lee et al., 2013). However, almost half of U.S. adults believe that humans were created, in their present form, at some point in the past 10,000 years (Gallup, 2012). As a specific example, less than 60% of Catholics believe that evolution occured, despite the church's explicit endorsement of that scientific theory (Pew Research Center, 2009). Just as we saw with Galileo, institutional doctrines do not always translate into individual religious belief. This personal aspect of religion—the beliefs one holds that may or may not match to officially endorsed doctrines—plays a prominent role in how religion and science influence the cultural landscape of the 21st century.

In addition to a gap between "official" religious stances and the beliefs held by the religious themselves, we also see divides between the consensus of scientists and beliefs held by the average person on issues where science and religion overlap. For example, there is very strong scientific agreement that not only have average global temperatures risen at an unprecedented rate over the past 100 years, but also that the major contributor to this is the release of greenhouse gases from human industry. Some 97% of peer-reviewed journal articles since the early 1990s point to man-made causes of global warming (Cook et al., 2013), whereas a recent report by the Intergovernmental Panel on Climate Change (IPCC) asserts a 95% certainty of the same (IPCC, 2014). For the lay public, however, things are different, with only around 60% of Americans believing that anthropogenic climate change is real. The numbers on evolution show even more of a scientist–public divide in the United States—only 32% of Americans agree that humans and other living things evolved as a result of natural processes, whereas 87% of scientists across all disciplines agree that we did (Pew Research Center, 2009). Some 42% of the American public believes that homosexuality is a choice people make, whereas scientists are united in the opinion that sexuality is determined by a complex interaction of genetic, prenatal environment, hormonal, and social factors (Pew Research Center, 2015). We do not see such a large scientist–public gap concerning scientific theories or findings for things such as general relativity, the principles of aerodynamics, or that the Earth moves around the sun (in the modern world, at least). These gaps, instead, tend to cluster around subjects in which one's religion and culture heavily influence beliefs and when subjects are more tangible and immediate.

For instance, on the issue of climate change, American evangelical Christians (such as Southern Baptists, Pentecostals, and many nondenominational churches) have been found to be less likely than nonevangelicals to believe that global warming is real, that it is caused primarily by human activities, and that it is causing serious harm (Smith & Leiserowitz, 2013). This sentiment is evidenced in the "Evangelical Declaration on Global Warming" put out by the Cornwall Alliance for the Stewardship of Creation (2009). This group claims not only that current climate change is part of a natural cycle that has little or nothing to do with human activities, but also that attempts to protect the environment, such as the placement of restrictions on carbon emissions, will be detrimental to economies around the world. This view is also often echoed by politicians, such as long-time Oklahoma Senator James Inhofe, an evangelical Christian who in 2012 published a book titled *The Greatest Hoax: How the Global Warming Conspiracy Threatens Your Future*. In an interview at the time of the book's release, Inhofe (2012) argued "the arrogance of people to think that we, human beings, would be able to change what He [meaning the Christian God] is doing in the climate is to me outrageous" (Tashman, 2012).

But this view is not a unanimous stance within the evangelical community, as several prominent Christian evangelical individuals and organizations are actively fighting to address climate change issues. Most notable is Sir John Houghton, an Oxford professor of atmospheric physics who played a leading role in developing the IPCC. In addition, the Evangelical Climate Initiative (ECI), which urges Christians to address anthropogenic climate change on both moral and scientific grounds, consists of more than 300 senior U.S. evangelical leaders and is endorsed by the National Association of Evangelicals, representing over 45,000 churches in the United States (Thompson, 2013). This ideological divide within the same general religious group seems to result primarily from two forces: differing biblical interpretations and funding sources.

In the Book of Genesis, specifically verses 26 to 28 in the first chapter, humans are granted "dominion" over the Earth and all life on it by the Judeo-Christian God. Differing interpretations of what this means have driven much of the split between evangelicals on the climate change issue. Some, like the ECI, see such dominion as indicating that humans should be good stewards of the Earth, caring for it and protecting it (what has been called "creation care"). Others, like the Cornwall Alliance, seem to interpret this more as meaning the Earth is ours to exploit as we will. Rick Santorum, a religiously conservative Catholic and Republican politician echoed the latter position by saying that "We were put on this Earth as creatures of God to have dominion over the Earth, to use it wisely and steward it wisely, but for our benefit not for the Earth's benefit" (Hooper, 2012). Much like Galileo found himself at odds with the Catholic Church of the early 1600s as a result of differences in how they interpreted scripture, these differences in reading can have marked impacts on attitudes, beliefs, and behavior in the real world. There is another, hidden aspect to this split, though. What is not mentioned in their "declaration" is that the primary funders of the Cornwall Alliance are the oil and gas giants ExxonMobil, Chevron, and a private foundation whose wealth comes from the same industry. Interestingly enough, oil and gas companies have also been the major contributors to Inhofe's campaign funds since the early 2000s, when he was elected chair of the Senate Committee on Environment and Public Works.

When it comes to the issue of climate change, evangelical Christians are not a united front. On the one hand, a feeling of obligation to God's creation motivates some evangelical leaders to support environmental protection and the notion that humans should act as stewards of the land. On the other hand, evangelicals who are not in favor of environmentalism may be motivated by oil and gas interests, rather than only because of the idea they were "given" dominion over the land. As such, although this may appear to be a debate between religious leaders and scientists, the truth is far more complex. This case briefly illustrates how the "personal" nature of religious belief can cause individuals to believe different things even when they share the same religious background, not unlike Galileo and the Catholic Church in the early 1600s. A person's beliefs, attitudes, and behaviors are influenced not merely by his or her interpretations of a sacred text, but by numerous background and social factors as well. This complexity, where religious beliefs are only one part of what influences behaviors and values, is common across so-called "culture war" issues. Although some people will choose to side with the scientific consensus (e.g., Sir John and the Evangelical Climate Institute), they will often do so not merely because "the data has spoken," but for multiple and sometimes complex reasons.

This idea—that a complex interplay of religious beliefs, sociocultural factors, politics, money, and scientific evidence leads to a person's position on a particular issue, which may or may not be the same as another person of the same religious background—repeats itself over and over in the modern world. For example, for almost a decade and a half, from the mid-1990s to the early 2010s, the U.S. federal government only provided sexual education funding to abstinence-only programs. This was despite mounds of scientific studies showing that such programs resulted in higher rates of sexually transmitted infections and teen pregnancy than comprehensive sexual education programs that taught about safer sex options as well. The abstinence-only promoters were not motivated by evidence of effectiveness, but instead by their religious beliefs that sexual activity prior to marriage was wrong and that comprehensive sexual education would only encourage such behavior. And just as with climate change and evolution, not only were these beliefs and programs at odds with scientifically derived data but also with a large percentage of Americans, some 60% to 80% of whom wanted comprehensive sex education programs in schools (Advocates for Youth, 2009).

These divides and schisms, which have shifted and morphed over time, are why anthropologist Eller (2008) argues that religions are best understood on a micro and not macro level. To take the example of Christianity, he argues that the diversity of thought, doctrines, and behaviors by people who consider themselves Christians almost renders the word useless. Instead, one must understand the "local Christianity" (or local Hinduism or local Islam) as being situated in a particular time and place as practiced by a certain group of people. As such, it does not make much sense to talk about some sort of war or conflict between science and religion, both because religion is not a unitary construct and because the vast majority of scientific findings do not come into conflict with people's religious beliefs. Further, as shown in this chapter, even when scientific theories or evidence do conflict with some people's beliefs, there is no guarantee that all people who are religious will feel the same way. Another lesson that these examples offer us, in the context of a text on critical thinking such as this, is that we should be wary of simplistic narratives related to religious identity and conflict, as these narratives might sometimes be little more than caricature.

CONCLUSIONS

Belief in a particular idea is a highly individual event, influenced by a plethora of factors. As has been stressed throughout this book, people rarely believe things for no reason at all. But sometimes they do believe things for the wrong reasons. As discussed extensively in Chapters 5 and 6, our brains and our world often conspire to make us believe things that are not demonstrably true. As we have seen, pseudoscientific beliefs are powerful, from the millions of people who believe they have been abducted by aliens to the billions of dollars spent each year on medical and psychological treatments that have been shown to be ineffective. In this chapter, you've seen a few examples of how religious beliefs can impact whether one engages in scientific skepticism or not, trusting empirical evidence to guide beliefs or relying on personal anecdotes or doctrine. You have also seen that there is not as much substance to the idea of a war of "science versus religion" as some insist there is, and that instead conflict between the two is often restricted to a few particular topics for a few particular types of believers.

We hope this allows our readers to appreciate the subtle and nuanced way that science and religion interact with each other and, in turn, influence our culture and world.

QUESTIONS FOR REFLECTION

1. *Do you think the so-called "culture wars" are largely a thing of the past, with religion having accepted the primacy of science? Or will there continue to be conflicts over the types of big questions discussed in this chapter?*
2. *What might religion offer society that cannot be (rather than is not currently) provided by science?*
3. *Is there a fundamental conflict between religion and science? How should governments and policy makers respond to this if there is? If there is not, how should those same entities respond to those claiming that there is?*
4. *In what other ways might the ascendancy of science in much popular discourse influence the evolution of religious views, besides those views increasingly accommodating scientific thinking?*
5. *Even among the religious, to what extent do you believe that sacred texts influence worldview? In other words, how much of one's worldview is drawn from one's religion versus the secular culture?*

REFERENCES

Advocates for Youth. (2009). *Comprehensive sex education: Research and results.* Retrieved from http://www.advocatesforyouth.org/publications/1487

Bellah, R. N. (2011). *Religion in human evolution: From the paleolithic to the axial age.* Cambridge, MA: Belknap Press.

BioLogos Foundation. (n.d.). *About the BioLogos foundation.* Retrieved from http://biologos.org/about

Cornwall Alliance. (2009). Evangelical declaration on global warming. Retrieved from http://www.cornwallalliance.org/2009/05/01/evangelical-declaration-on-global-warming

Cook, J., Nuccittelli, D., Green, S. A., Richardson, M., Winkler, B., Painting, R., . . . Skuce, A. (2013). Quantifying the consensus on anthropogenic global warming in the scientific literature. *Environmental Research Letters, 8*(2), 1–7.

Darwin, F. (Ed). (1887). *The life and letters of Charles Darwin, including an autobiographical chapter (Vol. 1).* London, UK: John Murray. Retrieved from J. van Whye (Ed.), *The complete works of Charles Darwin online* (http://darwin-online.org.uk).

Eller, D. (2008). *Atheism advanced: Further thoughts of a freethinker.* Cranford, NJ: American Atheist Press.

Fara, B. (2009). *Science: A four thousand year history.* Oxford, UK: Oxford University Press.

Gallup. (2012). *In U.S., 46% hold creationist view of human origins.* Retrieved from http://www.gallup.com/poll/155003/hold-creationist-view-human-origins.aspx

Gould, S. J. (1999). *Rocks of ages: Science and religion in the fullness of life*. New York, NY: Ballantine Books.

Haught, J. F. (1995). *Science and religion: From conflict to conversation*. Mahwah, NJ: Paulist Press.

Hooper, T. (2012). *Santorum and Gingrich dismiss climate change, vow to dismantle EPA*. Retrieved from http://www.coloradoindependent.com/111924/santorum-and-gingrich-dismiss-climate-change-vow-to-dismantle-the-epa

Inhofe, J. (2012). *Greatest hoax: How the global warming conspiracy threatens your future*. Long Beach, CA: WND Books.

Intergovernmental Panel on Climate Change. (2014). *Climate change 2014: Synthesis report. Contribution of working groups I, II and III to the fifth assessment report of the intergovernmental panel on climate change* [Core Writing Team, R. K. Pachauri & L. A. Meyer (Eds.)]. Geneva, Switzerland: Author.

James, W. (1902). *The varieties of religious experience*. New York, NY: Collier Books.

Koenig, H. G. (2008). *Medicine, religion, and health: Where science and spirituality meet*. West Conshohocken, PA: Templeton Press Foundation.

Lee, E., Tegmark, M., & Chita-Tegmark, M. (2013). *The MIT survey on science, religion and origins: The belief gap*. Retrieved from http://space.mit.edu/home/tegmark/survey/survey.pdf

Lindberg, D. C., & Numbers, R. L. (1987). Beyond war and peace: A reappraisal of the encounter between Christianity and science. *Perspectives on Science and Christian Faith, 39*(3), 140–149.

National Academy of Sciences & Institute of Medicine. (2008). *Science, evolution, and creationism*. Washington, DC: National Academies Press.

Noll, M. (1994). *The scandal of the evangelical mind*. Grand Rapids, MI: Eerdmans.

Pew Research Center. (2009). *Public praises science; scientists fault public, media*. Retrieved from http://www.people-press.org/2009/07/09/public-praises-science-scientists-fault-public-media

Pew Research Center. (2015). *Americans are still divided on why people are gay*. Retrieved from http://www.pewresearch.org/fact-tank/2015/03/06/americans-are-still-divided-on-why-people-are-gay

Polkinghorne, J. (1998). *Science and theology*. Minneapolis, MN: Fortress Press.

Scott. E. C. (1998). "Science and religion," "christian scholarship," and "theistic science." *Reports of the National Center for Science Education, 18*(2), 30–32.

Shermer, M. (2002). *Why people believe weird things*. New York, NY: Holt.

Smith, N., & Leiserowitz, A. (2013). American evangelicals and global warming. *Global Environmental Change, 23*(5), 1009–1017.

Tashman, B. (2012). James Inhofe says the Bible refutes climate change. *Right Wing Watch*. Retrieved from http://www.rightwingwatch.org/content/james-inhofe-says-bible-refutes-climate-change

Thompson, K. (2013). *Christian evangelicalism and climate change denial*. Retrieved from http://www.skepticink.com/gps/2013/11/20/christian-evangelicalism-and-climate-change-denial

White, A. D. (1896/1993). *A history of the warfare of science with theology in Christendom*. Amherst, NY: Prometheus Books.

CHAPTER 14

CONCLUSIONS AND RECOMMENDATIONS

We hope that you've enjoyed this excursion into critical thinking, science, psychology, and religion as much as we enjoyed writing it. More to the point, we hope we've provided you with some ideas and arguments that you can bring to bear in your own fields of interest. Because as much as some of us might work in the classroom, or in the laboratory, we shouldn't forget that we encounter true and false claims, and good and bad arguments, in all aspects of our lives. Intellectual curiosity and scientific skepticism are not tools that you dust off only when challenged in argument—they are instead the foundation of an epistemically (from *epistemology*, the area of philosophy concerned with the nature of knowledge) responsible life. What is an epistemically responsible life? It's one that is concerned with doing your best to believe things that are more likely to be true, and disbelieving the things that are more likely to be false—because you know that, in general, true beliefs lead to more efficient choices, in that they map onto the world more accurately than false beliefs do, whether those false beliefs are simple mistakes or something more serious, such as a deep-rooted prejudice.

Being skeptical in one's daily life means not being complacent about getting things right, and being aware of the sorts of all-too-common mistakes we're inclined to make, and which are discussed in this book. But being skeptical does not mean being negative, pessimistic, cynical, or any such thing (although you might of course be any or all of those things, regardless of whether you're a skeptic or not!). It means reserving judgment where evidence is incomplete, and reaching the conclusion best justified by the available evidence, whether or not that conclusion is comfortable or not, and no matter whether it disagrees with what you'd prefer to be true instead.

You know as well as we (the authors) do that critical thinking and a skeptical outlook aren't always welcomed by people we might end up talking or debating with. Few people like being told that they are wrong, or that they aren't thinking about an issue carefully enough. Furthermore, we have lived with some of our beliefs for so long that they might almost seem beyond questioning—axiomatic or obviously true in a way that makes critical interrogation difficult, if not impossible. What this should remind you of is that there are strong political and rhetorical dimensions to argumentation. People will be less inclined to listen and discuss things with an open mind if they feel attacked or belittled, so as much as we might occasionally feel that we are obviously right, it's important to remember that others might feel the same way—and that in any given case, it's certainly possible that it's we who are actually in the wrong!

To put that point more simply, some epistemic humility can go a long way to generating a productive discussion, meaning that all parties accept the possibility that they can learn something from the others. Of course, that's sometimes not the case, but if you care about fruitful dialogue, it's a useful starting point, especially when you don't know whether you share common ground or background assumptions with the people you're interacting with. But most important, a good skeptic always accommodates the possibility that she could herself be wrong, instead of treating her skepticism as license to assume the high ground in any given argument. It's always the evidence and arguments themselves that establish the high ground, not the identities or histories of those presenting the evidence or arguments. As Richard Feynman said, "I can live with doubt and uncertainty. I think it's much more interesting to live not knowing than to have answers that might be wrong" (p. 12).

THE IDEAL SKEPTIC

In this final chapter, we want to reinforce and put forth certain values that a good skeptic should aspire to (for a more humorous take, refer to Madison & Madison, 2011). Although none of us will likely be able to truly display all of these, we can at least strive to do so in our interactions with others. Given that many conflate the term *skeptic* with a sense of arrogance, one of the first and major qualities an ideal skeptic should have is *humility*. This is not to say that we should have a low view of ourselves, but instead that we learn to tolerate uncertainty and ambiguity and recognize the limits of our own knowledge. We must also strive to become comfortable with saying "I don't know" and be *introspective* enough to recognize that it is not a flaw, but a virtue to recognize what you do and do not know.

A good skeptic also keeps an *open mind*, understanding that we must be willing to change our beliefs (whether they concern scientific, moral, or social problems) in response to new evidence.

A major point of this book was to show not just that certain pseudoscientific beliefs are wrong, but also to show *why* people believe them. This was done to help foster *empathy*, another key trait of the ideal skeptic. Understanding someone else's point of view, even if it contradicts the evidence or your beliefs, is key to fostering productive communication, as described previously. Finally, the good skeptic strives to *apply their skepticism to all areas of life*, both external and internal. If you cannot take a critical look at your own beliefs or positions on certain issues, then you likely cannot be trusted to appropriately apply skepticism to those of other people.

ENCOURAGING CRITICAL THINKING VIA GRASSROOTS ACTIVISM

We all know how frustrating it can be to encounter irrationality and stubborn prejudice, so it's also to your own benefit to encourage the spread of critical thinking and scientific skepticism. Not only for preserving your good humor, but also because each of us is only one person in a global population of over 7 billion—meaning that there's ample opportunity for other people's mistakes in reasoning to negatively impact on you! So, how can you encourage good thinking habits in your personal lives, workplaces, and so forth?

The simplest and most obvious way is by setting an example in your own reasoning, through applying the principles expressed in this book and thereby showing others that although it's impossible to be a perfect thinker, it's relatively easy (for most of us) to become a better thinker than you currently are (but remember to not be an arrogant ass while doing so). So don't be afraid to ask for evidence when necessary, and to point out that the evidence provided isn't enough to substantiate the claims made. Be sure that when you're trying to persuade someone else of something, that you've presented the case as fairly and thoroughly as you can.

When engaging with others in debate, remember that debates occur in a context. Your opponent is rarely stupid, or irredeemably deluded—more often he simply has different motivations to yours, as well as access to a different data set (regardless of its quality relative to yours). So, to paraphrase Dennett (2013), we might usefully be reminded of the importance of applying Rapoport's rules when in argument. These rules were formulated by social psychologist Anatol Rapoport (the inventor of the tit-for-tat strategy in game theory), and they invite us (in Dennett's formulation) to argue as follows:

1. You should attempt to reexpress your target's position so clearly, vividly, and fairly that your target says, "Thanks, I wish I'd thought of putting it that way."
2. You should list any points of agreement (especially if they are not matters of general or widespread agreement).
3. You should mention anything you have learned from your target.
4. Only then are you permitted to say so much as a word of rebuttal or criticism.

One effect of following these rules is that your opponents in debate are instantly more likely to be receptive to your criticism than they would otherwise have been—even if they might never be sufficiently receptive. But you have done what you can in sending the clear signal that you're a fair and honest interlocutor, interested in arriving at the truth rather than simply defending a preexisting point of view.

Because, at the end of the day, our worth as skeptics is not vested in conclusions, but in the manner in which we reach conclusions. Skepticism is not about merely being right. Being right—if we are right—is the end product of a process and a method, not an excuse for some sanctimonious hectoring or feelings of superiority. Sometimes we need to remind ourselves of what that method looks like, and the steps in that process, to maximize our chances of reaching the correct conclusion. Focusing simply on the conclusions rather than the method can make us forget how often—and how easily—we can get things wrong.

As skeptics, we need to set an example in the domain of critical reasoning, and show others that regardless of authority or knowledge in any given discipline, there are common elements to all arguments, and that everybody can become an expert—or at least substantially more proficient—in how to deploy and critique evidence and arguments. That is part of what this book has been about. Because we believe that as humanism can be to ethics—a "woo-woo" free inspiration and guide for living a good life—skepticism should be for science, providing resources and examples of how to be a responsible believer, and of the importance of holding yourself responsible for what you believe.

RESOURCES

Besides the various texts and articles we used as references in writing this book, here is a brief selection of various other websites and organizations that you might find of interest and value. There is obviously a stupendously large number of possible resources we could list, so readers should please not take offense at finding that one of their favorites is missing here—these reflect both a desire not to overwhelm the reader and also the idiosyncratic preferences of your authors.

- The Science of Everyday Thinking—a free online course offered by the University of Queensland, Australia, on the edX platform: www.edx.org/course/science-everyday-thinking-uqx-think101x
- Skepticism 101—"a comprehensive, free repository of resources for teaching students how to think skeptically": www.skeptic.com/skepticism-101

- The Committee for Skeptical Inquiry (CSI—a program of the Center for Inquiry) Resource collection—a database of articles, blogs, forums, and publications related to skepticism: www.csicop.org/resources
- If you're interested in getting involved in skeptical activities, the Center for Inquiry has a large international network that you can learn about here: http://www.centerforinquiry.net/outreach/cfi_network. For example, in South Africa, one of your authors (JR) is the Chair of the Free Society Institute, dedicated to promoting humanism and scientific skepticism in that part of the world (http://fsi.org.za)

REFERENCES

Dennett, D. (2013). *Intuition pumps and other tools for thinking*. New York, NY: W. W. Norton.

Feynman, R. P. (1999). *The pleasure of finding things out*. New York, NY: Perseus Books.

Madison, M., & Madison, S. (2011). *Who are the skeptic illuminati?* Retrieved from http://www.skepticismandethics.com/2011/10/who-are-skeptic-illuminati.html

AFTERWORD: SCIENCE AND HUMILITY

If you enter the phrase "scientists are arrogant" into the Google search engine, as I did on September 27, 2015, when writing this Afterword, you'll obtain over 4,500 web hits. By the time you're reading this book, the number will very likely be higher. One website quotes an unnamed professor specializing in social sciences as saying that "Scientists are arrogant. When they move up in the hierarchy, their ego also goes up. They think no one can question them" (www.dnaindia.com/mumbai/report-arrogant-exclusivist-politicized-tiss-professor-on- scientists-2050617). Another blog writer opined that "Simply put, scientists are arrogant. Those of you who've spent time in academia are probably wondering why I waste time stating something so obvious" (Kulkarni, 2010). This view is mirrored in much of popular culture. For example, one prototypical depiction of the scientist in Hollywood movies is that of the mad genius who is ambitious and willing to flout ethical principles in pursuit of fame and glory (Weingart, Muhl, & Pansegrau, 2003).

THE PERSONALITY OF THE SCIENTIST

In fairness, there's a kernel of truth in the mainstream portrayal of scientists. As anyone who's spent more than a few days in an academic setting knows, some scientists can be dogmatic, pigheaded, and close-minded. The provocative documentary, *A Flock of Dodos* (Olson, Janata, Carlisle, & Miller, 2006), which contrasts the arguments of the proponents of evolutionary theorists with those of intelligent design theory, makes light of the condescending and high-handed communication styles of many biological scientists while acknowledging that they are correct about the science of Darwinian natural selection. After watching this documentary, one may conclude with some justification that we scientists are at times our own worst enemies when it comes to communicating the fruits of our knowledge to the general public. When considering the impression we make on laypersons, we'd be well-advised to follow the advice of one author, who titled his book *Don't Be Such a Scientist* (Olson, 2010).

Indeed, personality data reveal that scientists tend to be more arrogant, interpersonally dominant, self-confident, hostile, and ambitious than the average person; these traits appear to be especially marked among the most creative scientists (Feist, 2006). In retrospect, these findings should perhaps not be terribly surprising. After all, science is a relentless pursuit of the truth, people's feelings be damned. Science is a contact sport (Schneider, 2009) and is not for the thin-skinned or faint of heart. Hence, it's not entirely unexpected that many of the most successful scientists have outsize egos. They must be willing to advocate for their preferred hypotheses despite forceful, even withering, criticism. In this respect, a tad bit of arrogance, or at least self-assurance in the face of fierce opprobrium from colleagues, probably comes in handy as a scientist.

DISTINGUISHING *SCIENCE* FROM *SCIENTISTS*

Still, it's essential to distinguish *scientists* from *science* (Esolen, 2012). Scientists, being human, are inherently subject to insecurities, petty rivalries, and back-biting (see Watson [1968] for a classic illustration that is a "must read" for all students of science). Moreover, data demonstrate that scientists are at least as susceptible to confirmation bias—the deeply entrenched propensity to seek out evidence consistent with what we believe, and to deny, dismiss, or distort evidence that isn't (Nickerson, 1998)—as is everybody else (Mahoney & DeMonbreun, 1977). Nevertheless, although scientists can certainly be arrogant, *science itself is actually a prescription for humility* (McFall, 1996), a crucial point that the authors of this superb book wisely remind us of in Chapter 14.

This assertion may seem paradoxical or contradictory, but it isn't. Although scientists' personal biases may at times lead them to embrace poorly supported conclusions, the fundamental safeguards of science—including rigorous research methodology, an insistence on replication of one's findings, and peer review—ultimately hold their feet to the fire and force them to put up or shut up. Scientists push hard for their favored views, but the scientific community pushes back by insisting on methodological rigor and, when necessary, more compelling evidence for their claims (Neuringer, 1991). And when the data that are invoked to buttress scientists' positions are deeply suspect, the scientific community pushes back hard, often by criticizing its colleagues harshly in published forums and at academic conferences. As the authors of this book remind us, this epistemic humility—the awareness that one might be incorrect—is a cardinal feature of skepticism. It is also what distinguishes science from many other ways of knowing, such as fundamentalist religion and partisan politics.

Let's consider the peer-review process as a core feature of science. As a psychological researcher who's had his fair share of article rejections, I can assure you that peer review can be a profoundly humbling and ego-bruising experience. Upon reading negative reviews of one's submitted manuscript, one looks on helplessly as one' hypotheses, methodology, and conclusions are ripped apart mercilessly by one's colleagues. Yet, I can also assure you that peer review, although hardly perfect (see Kuehn, 2013; Peters & Ceci, 1982), is an invaluable bulwark against human error. Over the years, peer reviewers have caught mistakes and ambiguities in my writing, thinking, and analyses, and, more broadly, have enhanced the quality of most of my manuscripts. Although one's initial reaction to negative peer reviews may often be defensiveness or outright hostility, this response is rarely warranted. As psychologist Benton Underwood (1957) wrote:

> The rejection of my own manuscripts has a sordid aftermath: (a) one day of depression; (b) one day of utter contempt for the editor and his accomplices; (c) one day of decrying the conspiracy against letting Truth be published; (d) one day of fretful ideas about changing my profession; and (e) one day of re-evaluating the manuscript in view of the editor's comments followed by the conclusion that I was lucky it wasn't accepted! (p. 87)

The humility of science is misunderstood by many critics. Two weeks ago, I read a draft of an eloquent essay by an author who has long been dubious of the need for evidence-based practice in my own field of clinical psychology. Evidence-based practice emphasizes the use of the best available research data to guide one's selection of psychological treatments and assessment techniques, in conjunction with client's preferences and values (Lilienfeld, Ritschel, Lynn, Cautin, & Latzman, 2013; Spring, 2007). The author contended, in essence, that evidence-based practice undermines clinicians' freedom in selecting interventions and devalues the role of clinical intuition.

Yet psychologists have known for over 60 years that when it comes to integrating information from multiple sources of evidence to reach accurate decisions, statistical equations derived from extant data tend to be superior to subjective judgments; with few exceptions, it's better to "use the formula" rather than to "use our heads" (Meehl, 1954). Physician Mark Crislip (see Hall, 2015) has written that the most dangerous words in medicine are "in my experience." Indeed, the sordid histories of psychiatry and psychology teach us that many or most of the most disastrously harmful interventions for mental illness, such as prefrontal lobotomy, recovered memory techniques, coercive attachment therapies (such as rebirthing and holding therapies), and facilitated communication, were largely borne out of clinical intuition and subjective observations of improvement in the absence of rigorous controlled data. Furthermore, the author's essay overlooks the critical point that although clinical intuition is sometimes correct, it is frequently incorrect. Only science can allow us to tell the difference

(Dawes, 1994). In this respect, a scientific approach, not one informed exclusively by gut hunches or clinical experience, is ultimately the most humane way of treating patients, because it minimizes the risk of error (Dawes, Faust, & Meehl, 1989).

THE ESSENCE OF SCIENCE

When viewed in this light, we can better appreciate what science is all about. Science, at its core, is a finely honed set of safeguards designed to compensate for biases, especially confirmation bias (Lilienfeld, 2010). Put a bit differently, science is an armamentarium of tools that scientists have developed and refined over the centuries to prevent themselves from fooling themselves. Nobel prize-winning physicist Richard Feynman (1985) got it exactly right when he described science as a process of bending over backward to prove ourselves wrong. Social psychologists Carol Tavris and Elliot Aronson (2007) similarly referred to science as a method of "arrogance control" (p. 109). Science keeps researchers' epistemic arrogance—although not necessarily their personal arrogance—in check by subjecting their claims to careful scrutiny.

If this analysis is on target, the best scientists are not those who are immune from confirmation bias. No one is. Instead, the best scientists are those who are cognizant of their propensities toward confirmation bias and avail themselves of the protections afforded by science to counteract them. In the parlance of contemporary social cognition, the best scientists are characterized by a small *bias blind spot*. Bias blind spot is the pervasive tendency to be aware of biases, including confirmation bias, in other people but not in ourselves (Pronin, Lin, & Ross, 2002). This meta-bias—a bias about biases—can be an obstacle to scientific truth-seeking, because researchers who regard themselves as immune to bias may neglect to implement procedural safeguards against error, such as blinded research designs, sophisticated statistical analyses, and the need to replicate one's findings before promoting them zealously to their colleagues or the general public. As astronomer Carl Sagan and his wife, science writer and producer Ann Druyan, observed, "Science ... is forever whispering in our ears, 'Remember, you're very new at this. You might be mistaken. You've been wrong before'" (Sagan, 1995, p. 39).

Charles Darwin was a great scientist not only because he was remarkably brilliant and creative, but also because he was keenly aware of his own susceptibility toward confirmation bias, including selective memory. He wrote:

> I had also, during many years, followed a golden rule, namely that whenever published fact, a new observation of thought came

> across me, which was opposed to my general results, to make a memorandum of it without fail and at once; for I had found by experience that such facts and thoughts were far more apt to escape from the memory than favourable ones. (Barlow, 1958, p. 123)

What a wonderful quote! Darwin is acknowledging his own propensity to selectively recall his hits and forget his misses, a hallmark feature of confirmation bias noted by scholars at least as far back as Sir Francis Bacon (see also Jevons, 1874). What's more he even develops a simple procedure to debias himself against this error, namely, promptly jotting down any observations that appeared to contradict his theory of natural selection before he conveniently forgot about them. Darwin had a small bias blind spot.

SCIENCE AS A SELF-CORRECTING ENTERPRISE

As a consequence of its continual efforts to root out errors, scientific knowledge is necessarily provisional. As the authors note in Chapter 2, some scientific knowledge, such as the knowledge that life forms evolve through a process of random mutation and natural selection, is so well-established that it can be regarded as well established; yet all scientific conclusions could in principle be overturned by fresh evidence. As a consequence, a key feature of science is its capacity for self-correction (Lilienfeld, Lynn, & Lohr, 2014). In this respect, such widely used terms as "proof" or "settled science" are antithetical to how science works (Lilienfeld et al., 2015; McComas, 1996).

Still, we shouldn't be pollyannaish. Science, being conducted by human beings, doesn't always operate this smoothly. The self-correction of science can be short-circuited or delayed by close-mindedness. For example, although German researcher and explorer Alfred Wegener proposed the theory of continental drift in 1912, his views were resisted and even mocked for five decades until they gained widespread acceptance in the 1960s (Frankel, 1979). More broadly, science often self-corrects far more slowly than we'd like (Ioannidis, 2012). This is especially the case when it comes to the application of scientific findings to real-world settings in such domains as medicine and psychology. For example, even though percutaneous coronary intervention (stenting) for stable coronary artery disease and hormone therapy for postmenopausal women were discredited by rigorous controlled trials, they continued to be administered for many years and endorsed by defenders in the published literature (Prasad, Cifu, & Ioannidis, 2012). Facilitated communication for autism, which has been thoroughly debunked in carefully controlled scientific trials (see Chapter 12), continues to be implemented in many quarters and may even be experiencing a comeback. Even a few academicians at major universities continue to support its use despite overwhelmingly negative research evidence (Lilienfeld, Marshall, Todd, & Shane, 2014).

The beauty of science, however, is that it's an engine of self-criticism. Scientists themselves have been at the forefront of concerted efforts to criticize certain domains of science for their unduly slow self-correction, and they are busily at work attempting to improve the quality of research and clinical practice. Despite their best efforts, however, some medical and mental health professionals will continue to administer unsupported or refuted techniques in their everyday practices. Hope—and pseudoscience—springs eternal.

SCIENCE VERSUS PSEUDOSCIENCE

Given everything I've discussed, how should we conceptualize pseudoscience? The boundaries between science and pseudoscience are fuzzy (Pigliucci & Boudry, 2013). In part, that's because many of the classic indicators of pseudoscience, such as exaggerated claims, excessive reliance on anecdotal evidence ("anecdata"), and cherry-picking of positive outcomes (see Chapter 3), surely fall on a continuum from occasional to extremely frequent. Also, scientists themselves are by no means immune to these unsavory practices from time to time.

Still, that doesn't mean that we can't meaningfully distinguish clear-cut cases of science from clear-cut cases of pseudoscience. As psychophysicist S. S. Stephens noted, the fact that day and night shade imperceptibly into each other (think of dawn and dusk) doesn't imply that we can't distinguish day from night (see Leahey & Leahey, 1983). With that point in mind, we can now understand what ties together such diverse disciplines as astrology, UFOlogy, parapsychology, and cryptozoology, at least as they're typically practiced: *Pseudosciences tend to lack the systematic safeguards against confirmation bias that characterize most mature sciences*. Hence, they do not possess the epistemic humility that one finds in science, at least in science at its best. In contrast to most sciences, pseudosciences usually avoid subjecting their cherished claims to severe tests; that is, studies that place their theories at grave risk of refutation (Popper, 1959). Adapting the words of Richard Feynman (1985), pseudosciences generally bend over backward to prove themselves right, rather than to prove themselves wrong.

I close with two caveats. First, we should not confuse epistemic humility with personal humility. Just as many purveyors of science are personally arrogant, many advocates of pseudoscientific claims are personally humble. Although I'm not aware of any systematic research comparing these two "flavors" of humility, I suspect that they are largely or entirely distinct. Second, we all have our blind spots. I suspect that most or all of us—myself included—hold certain utterly false beliefs with utmost conviction, especially in cases in which we have a deep personal or emotional commitment to a point of view. If I'm right, most or all of us harbor certain "pockets" of epistemic arrogance.

This point should remind us that we need to be patient with and tolerant of people who hold poorly supported beliefs, because we've sometimes walked in the same shoes ourselves without realizing it. The ideal skeptic doesn't condemn or deride others. Instead, he or she listens to others with an open mind, examines their assertions as impartially as possible, asks for convincing evidence, and tactfully but firmly raises pointed questions when such evidence isn't forthcoming. Returning full circle to this book's Preface, we should all bear in mind what skeptic Michael Shermer (1994) termed Spinoza's dictum: "I have made a ceaseless effort not to ridicule, not to bewail, not to scorn human actions, but to understand them" (Spinoza, 1667).

Scott O. Lilienfeld, PhD
Department of Psychology
Emory University
Atlanta, Georgia

REFERENCES

Barlow, N. (1958). *The autobiography of Charles Darwin, 1809–1882: With original omissions restored; edited with appendix and notes by his grand-daughter Nora Barlow*. London, UK: Collins.

Dawes, R. M. (1994). *House of cards: Psychology and psychotherapy built on myth*. New York, NY: Free Press.

Dawes, R. M., Faust, D., & Meehl, P. E. (1989). Clinical versus actuarial judgment. *Science, 243*, 1668–1674.

Esolen, A. (2012, January 26). The humility of science, the arrogance of scientists. *Crisis Magazine*. Retrieved from http://www.crisismagazine.com/2012/the-humility-of-science-the-arrogance-of-scientists

Feist, G. J. (2006). *The psychology of science and the origins of the scientific mind*. New Haven, CT: Yale University Press.

Feynman, R. P. (1985). *Surely you're joking, Mr. Feynman: Adventures of a curious character*. New York, NY: W. W. Norton.

Frankel, H. (1979). The career of continental drift theory: An application of Imre Lakatos' analysis of scientific growth to the rise of drift theory. *Studies in History and Philosophy of Science Part A, 10*, 21–66.

Hall, H. (2015). Evidence: "It worked for my Aunt Tillie" is not enough. *Skeptic, 20*(3), 7–8.

Ioannidis, J. P. (2012). Why science is not necessarily self-correcting. *Perspectives on Psychological Science, 7*, 645–654.

Jevons, W. S. (1874). *The principles of science*. New York, NY: Dover.

Kuehn, B. M. (2013). Striving for a more perfect peer review: Editors confront strengths, flaws of biomedical literature. *Journal of the American Medical Association, 310*, 1781–1783.

Leahey, T. H., & Leahey, G. E. (1983). *Psychology's occult doubles: Psychology and the problem of pseudoscience*. Chicago, IL: Nelson–Hall.

Lilienfeld, S. O. (2010). Can psychology become a science? *Personality and Individual Differences*, *49*, 281–288.

Lilienfeld, S. O., Lynn, S. J., & Lohr, J. M. (Eds.). (2014). *Science and pseudoscience in clinical psychology* (2nd ed.). New York, NY: Guilford Press.

Lilienfeld, S. O., Marshall, J., Todd, J. T., & Shane, H. C. (2014). The persistence of fad interventions in the face of negative scientific evidence: Facilitated communication for autism as a case example. *Evidence-Based Communication Assessment and Intervention*, *8*, 62–101.

Lilienfeld, S. O., Ritschel, L. A., Lynn, S. J., Cautin, R. L., & Latzman, R. D. (2013). Why many clinical psychologists are resistant to evidence-based practice: Root causes and constructive remedies. *Clinical Psychology Review*, *33*, 883–900.

Lilienfeld, S. O., Sauvigné, K. C., Lynn, S. J., Cautin, R. L., Latzman, R. D., & Waldman, I. D. (2015). Fifty psychological and psychiatric terms to avoid: A list of inaccurate, misleading, misused, ambiguous, and logically confused words and phrases. *Frontiers in Psychology*, *6*, 1100.

Mahoney, M. J., & DeMonbreun, B. G. (1977). Psychology of the scientist: An analysis of problem-solving bias. *Cognitive Therapy and Research*, *1*, 229–238.

McComas, W. F. (1996). Ten myths of science: Reexamining what we think we know about the nature of science. *Scholarship in Science and Mathematics*, *96*, 10–16.

McFall, R. M. (1996). Making psychology incorruptible. *Applied and Preventive Psychology*, *5*, 9–16.

Meehl, P. E. (1954). *Clinical versus statistical prediction: A theoretical analysis and a review of the evidence*. Minneapolis, MN: University of Minnesota Press.

Neuringer, A. (1991). Humble behaviorism. *Behavior Analyst*, *14*, 1–13.

Nickerson, R. S. (1998). Confirmation bias: A ubiquitous phenomenon in many guises. *Review of General Psychology*, *2*, 175–220.

Olson, R. (2010). *Don't be such a scientist: Talking substance in an age of style*. Washington, DC: Island Press.

Olson, R. (Director), Janata, J., Carlisle, T., & Miller, S. (Producers). (2006). *Flock of dodos: The evolution-intelligent design circus* [Documentary Film]. United States: Documentary Educational Resources.

Peters, D. P., & Ceci, S. J. (1982). Peer-review practices of psychological journals: The fate of published articles, submitted again. *Behavioral and Brain Sciences*, *5*, 187–195.

Pigliucci, M., & Boudry, M. (Eds.). (2013). *Philosophy of pseudoscience: Reconsidering the demarcation problem*. Chicago, IL: University of Chicago Press.

Popper, K. R. (1959). *The logic of scientific discovery*. New York, NY: Routledge Classics.

Prasad, V., Cifu, A., & Ioannidis, J. P. (2012). Reversals of established medical practices: Evidence to abandon ship. *Journal of the American Medical Association*, *307*, 37–38.

Pronin, E., Lin, D. Y., & Ross, L. (2002). The bias blind spot: Perceptions of bias in self versus others. *Personality and Social Psychology Bulletin*, *28*, 369–381.

Sagan, C. (1995). *The demon-haunted world: Science as a candle in the dark*. New York, NY: Random House.

Schneider, S. (2009). *Science as a contact sport: Inside the battle to save Earth's climate*. Washington, DC: National Geographic.

Shermer, M. (1994). How thinking goes wrong. *Skeptic*, *2*(3), 42–49.

Spinoza, B. (1667). *Tractatus Politicus*. Edited with an Introduction by R. H. M. Elwes. (A. H. Gosset, 1883, Trans.). Chicago, IL: Contemporary Books.

Spring, B. (2007). Evidence-based practice in clinical psychology: What it is, why it matters; what you need to know. *Journal of Clinical Psychology, 63*, 611–631.

Tavris, C., & Aronson, E. (2007). *Mistakes were made (but not by me): Why we justify foolish beliefs, bad decisions, and hurtful actions*. Boston, MA: Houghton-Mifflin.

Underwood, B. J. (1957). *Psychological research*. East Norwalk, CT: Apple-Century-Crofts.

Watson, J. D. (1968). *The double helix: A personal account of the discovery of the structure of DNA*. New York, NY: Atheneum.

Weingart, P., Muhl, C., & Pansegrau, P. (2003). Of power maniacs and unethical geniuses: Science and scientists in fiction film. *Public Understanding of Science, 12*, 279–287.

INDEX